Teaching and Learning Difficulties

Cross-curricular perspectives

Peter Westwood

Teaching and Learning Difficulties

Cross-curricular perspectives

Peter Westwood

ACER Press

First published 2006
by ACER Press
Australian Council *for* Educational Research Ltd
19 Prospect Hill Road, Camberwell, Victoria, 3124

Edited by Emma Driver
Designed by Polar Design Pty Ltd
Cover illustration by Sarn Potter Graphics
Typeset by Ian Thatcher
Printed by Shannon Books

National Library of Australia Cataloguing-in-Publication data:

Westwood, Peter S. (Peter Stuart), 1936- .
Teaching and learning difficulties : cross-curricular
perspectives.
   Includes index.
   ISBN-13: 978 0 86431 493 2.
   ISBN-10: 0 86431 493 0.

   1. Learning. 2. Effective teaching. 3. Education -
Curricula. 4. Learning disabilities. 5. Literacy - Study
and teaching. 6. Numeracy - Study and teaching. I. Title.

371.9043

Visit our website: www.acerpress.com.au

Contents

For Chan Wing Yan (Carol)

Preface

This book has been written as a companion volume to my text *Learning and learning difficulties* (2004). Here, I have attempted to explore in detail the many different teaching approaches available for use, describing their potential advantages and disadvantages. In particular, I have identified aspects of teaching approaches that may directly or indirectly cause students to have learning problems. I concur completely with Farkota's (2005) belief that many cases of learning difficulty can be traced to inappropriate or insufficient teaching, rather than to deficiencies in the students.

It has not been my intention to recommend one particular method as superior to all others for achieving all types of educational objective. Purdie and Ellis (2005) are right in suggesting that no single teaching method can possibly be appropriate for bringing about all types of learning. A teaching approach should be selected because of its goodness of fit for the type of learning involved in a lesson and for the learning characteristics of the students in that class.

In this book, teaching approaches have been categorised as belonging somewhere on a continuum between 'teacher-directedness' and 'student-centredness' in their emphasis. But as I point out later, such categories are misleading, because most approaches contain elements of both teacher direction and student-centredness. Lessons are rarely wholly teacher-centred or wholly student-centred, and effective teaching requires that an appropriate balance be achieved between the two.

An unusual feature of this book is that I have included coverage of learning difficulties and teaching methods in relation not only to basic academic skills (literacy and numeracy) but also to specific subject areas such as science, social studies, history, geography and environmental education. Potential causes of learning difficulty are discussed within the context of these subjects.

I have deliberately used as my primary source international literature on curriculum, learning difficulties and teaching methods, particularly from Britain, Australia, New Zealand and the US. It has always irritated me greatly over the years that professional literature in the field of learning difficulties tends to be very parochial in its perspective, dealing often with minutiae of local policies and practices in a particular country but failing to see the bigger picture in which key issues in teaching and learning are identical across all countries. I hope this text helps delineate the bigger picture by taking a cross-curricular and international perspective.

In this text I have introduced many references and resources that can be located online. This has been done to help any reader who wishes to investigate issues in greater detail but does not have easy access to an academic library. A disadvantage may be that certain websites may eventually disappear; for that I apologise in advance.

My sincere thanks to Emma Driver for applying her highly efficient editorial skills in preparing this manuscript for print, and to Lam Hiu Chi (Denise) and Mona Wong for technical assistance.

PETER WESTWOOD

Curriculum, teaching methods and learning difficulties

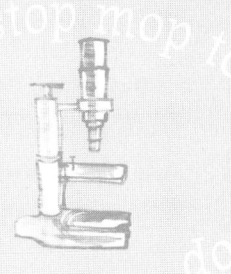

> The one point on which academic agreement can be said to exist is that the vast bulk of the problems associated with student learning can be directly related back to the nature of the curriculum or the method of teaching, and are not due to any lack of requisite intelligence or innate ability on the part of our students (Farkota, 2005, p.10).

In the quotation above, Farkota (2005) has highlighted the important but often overlooked fact that many problems in learning are not due to intrinsic causes such as lack of cognitive ability, perceptual impairment, learning disability, deficiencies in memory and attention or poor motivation. Instead, difficulties are caused or exacerbated by ineffective or insufficient teaching, or by curriculum content that is in some way inappropriate (Kershner, 2000; Robertson, Hamill & Hewitt, 1994). This is not a new perspective; many years ago Lembo (1971, p.7) wrote:

> While there are many complex factors, physical, psychological, economic and sociological, which account for each child's school performance, the basic cause of failure is the schooling process itself. Students do not enter school as failures. When students 'fail' it is the practices which teachers and administrators individually and collectively employ that are at fault.

This is not to suggest that certain characteristics of students do not predispose them to encounter difficulties in learning. Features such as sensory impairment, low intelligence, developmental delay, lack of support from home, frequent absences and so forth can naturally cause difficulties. Some of these factors will be described later in the context of learning particular school subjects. But Hattie (2003) has observed that although learner characteristics account for some 50 per cent of the variance in school achievement, another 30 per cent is accounted for by the quality of the teaching they receive. Very little can be done to modify most characteristics of learners (such as home background, poverty, health, disability, intelligence), but, in contrast, quality of teaching can be improved with obvious benefits for all students.

Despite awareness that extrinsic factors such as curriculum content and teaching methods significantly influence learning, there remains a tendency on the part of educators to seek explanations for learning difficulties in terms of so-called 'deficits' within the learner (Croll & Moses, 2000; McLaren, 2003). Under this model, remedial intervention involves prescribed exercises to reduce or eliminate deficits. Taking this doubtful diagnostic-prescriptive path has led in the past to some students being referred to remedial programs focusing on 'psychological processes', rather than addressing directly the problematic academic subject or skill. For example, some years ago when this writer was

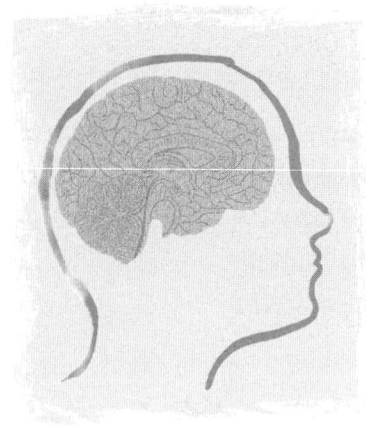

teaching students in a special class in primary school, it was believed that learning difficulties in reading were caused chiefly by problems in visual perception. As a result, some students with reading difficulties spent many hours working on material such as the *Frostig program for development of visual perception* (Frostig & Horne, 1964). Alternatively, a student may have been diagnosed as having problems in one or more 'psycholinguistic processes' such as auditory sequential memory or visual sequential memory. These students might then have found themselves involved in intensive psycholinguistic training to remedy these weaker

processes (Kirk & Kirk, 1971). Unfortunately, research studies have shown these process training programs — and others related to modality preferences, sensory integration and learning styles — to be ineffective in helping students achieve better results in academic learning (Cole & Chan, 1990; Howell, 1995; Swanson, 2000a). Worse still, the time students devoted to such unproductive programs was time lost to working more intensively on the weak academic subject.

Rather than looking for deficits within a student it is usually much more effective to investigate extrinsic factors such as the nature and relevance of the curriculum, and the quality and amount of instruction the student is receiving. These variables are more amenable to modification, and therefore improvement, than are factors within the student. Identifying how best to overcome learning difficulties involves finding the most significant factors that can be manipulated within the learning environment.

Curriculum as a source of difficulty

Robertson, Hamill and Hewitt (1994) suggested that the curriculum itself can contribute to difficulties in learning by:

- Presenting concepts that are pitched too far ahead of the cognitive level of the students.
- Moving forward too quickly compared to students' rate of learning.
- Using topics and materials that are of no interest or relevance to students of that age.
- Overloading the program with too much content.

In later chapters, reference will be made again to problems that arise when students are required to encounter concepts and ideas too far beyond their cognitive or affective level of development, for example in science, social studies and mathematics. Such programs are frustrating for students (and probably for their teachers too) because the students cannot acquire a deep understanding of the material. Instead, they adopt a superficial rote memorisation approach. Teachers attempting to teach higher-order concepts and abstractions beyond students' current cognitive capacity will always cause problems in learning. Such teaching will also have a negative impact on interest and motivation (Leiding, 2002). Van Kraayenoord (2002) has pointed out that students quickly become disengaged from a curriculum that does not connect with their lives, interests and prior knowledge. Elliott and Garnett (1994, p.6) described such students as 'curriculum disabled'.

Teaching methods as a source of difficulty

Perhaps even more than curriculum content, the teaching methods and resources used by a teacher can contribute significantly to learning problems (Robertson, Hamill & Hewitt, 1994). For example, a teacher may:

- Use instructional strategies that are not suited to the ways in which students learn most effectively.

- Adopt an approach that is too unstructured and informal for some students, or unsuitable for achieving certain learning objectives.

- Provide too few practical activities to engage and hold students' attention.

- Create insufficient opportunities for students to acquire knowledge and skills to mastery level before moving on to new topics.

- Review and revise previously taught material too infrequently.

- Make too little use of explicit teaching for information and strategies.

- Overestimate students' ability to learn independently.

- Provide too little feedback to students.

- Rely too much on a textbook approach to learning.

- Make too little use of visual aids, concrete materials, and information and communication technology (ICT).

- Manage time inefficiently, often leaving tasks unfinished or having 'dead' spots in a lesson.

- Fail to accommodate educationally relevant differences among learners.

- Communicate poorly with the students when instructing or when asking or answering questions.

- Talk too much.
- Listen to students too little.

All the factors above may cause or exacerbate learning difficulties. Students who begin to have difficulties soon lose confidence and may give up any attempt to improve. They develop what Brophy (1998) refers to as the 'Failure Syndrome', characterised by poor self-esteem, reduced motivation and weakened faith in their own self-efficacy. Teachers may regard them as 'problem learners' and lower their expectations regarding the students' potential.

If the points above identify poor teaching leading to learning difficulties, what is good teaching? And how does good teaching minimise learning problems? Two sources of information are available to help answer these questions — students' perceptions of teachers, and classroom research.

Students' views of good teaching

According to Batten, Marland and Khamis (1993), Morgan and Morris (1999) and McBer (2000), students tend to describe a 'good' teacher as one who:

- Helps you with your work.
- Takes time to explain things.
- Explains clearly so you can understand.
- Knows what he or she is talking about.
- Likes teaching his or her subject.
- Tells you how you are doing.
- Makes lessons enjoyable.
- Is friendly and easy to get along with.
- Is fair and straightforward.
- Cares about you, understands you and encourages you.
- Makes you feel clever.
- Is always ready to listen; allows you to have your say.
- Doesn't give up on you; makes allowances.
- Has a sense of humour.
- Controls the class well.
- Treats people equally.
- Is forgiving.

It is interesting to note that students' perceptions of 'good teaching' and a 'good teacher' include both an instructional skills dimension and a humanistic, empathic and caring dimension; students perceive both to be important in helping them learn. There are some interesting parallels between students' views and the objective findings from classroom research into teacher effectiveness.

The 'effective teacher' model

The classroom research on effective teaching examined the methods, strategies and characteristics of teachers when working with students, and then related these factors to the achievement outcomes for students (Jacobsen, Eggen & Kauchak, 2002; Killen, 1998; McBer, 2000). The results suggest that effective teachers exhibit the following characteristics.

Effective teachers tend to:

- Have well-managed classrooms.
- Provide students with the maximum opportunity to learn.
- Maintain an academic focus.
- Have high, rather than low, expectations of what students can achieve.
- Are businesslike and work-oriented.
- Show enthusiasm.
- Use strategies to keep students on task, motivated and productive.
- Impose structure on the content to be covered.
- Present new material in a step-by-step manner.
- Employ direct (explicit) teaching procedures.
- Use clear instructions and explanations.
- Use a variety of teaching styles and resources.
- Frequently demonstrate appropriate task-approach strategies.
- Monitor closely what students are doing during the lesson.
- Adjust instruction to individual needs, and reteach where necessary.
- Provide frequent feedback to students.
- Use high rates of questioning to involve students and to check for understanding.
- Spend significant amounts of time in interactive whole-class teaching, but also use group work and partner activities when appropriate.

In the list above it is pertinent to note that there are not many features of what might be described as an 'informal student-centred approach', with students encouraged to learn for themselves. Instead the list identifies mainly instructional, management and supportive strategies that effective teachers use to help sustain high achievement in their classes and minimise learning failure. What is perhaps most interesting is that the list very clearly shows effective teachers to be very much focused on students, their learning and their wellbeing. To that extent, effective teachers themselves are 'student-centred' in the way that they operate in the classroom.

It is sometimes suggested that what is represented in the effective teaching model is a 'behavioural' perspective on instruction, with an authoritarian and 'dehumanised' teacher controlling the learning situation. Students respond, are in some way rewarded or reinforced, and learning takes place. This is a very distorted view of effective teachers and teaching, as will be shown in the next chapter. Effective teaching involves a great deal of social and verbal interaction between teacher and students. The research indicates that effective teachers are actually warm, concerned and flexible in their approach (Wilen et al., 2000).

Harris (1998), when summarising main conclusions from classroom research, identified the following points:

- Students learn most in classrooms that are well-managed and provide clear structure and goals.
- Students who spend much of their time being instructed by teachers and working under direct supervision from teachers make better progress than students who are expected to learn more independently.
- Clear presentations, explanations, questioning and feedback (active teaching) are related to positive student learning.
- The pace of a lesson influences participation and learning.
- Time spent in study (academic engaged time) is a very important predictor of achievement.
- The teaching environment (including classroom climate and social interactions) influences learning.
- Different approaches to teaching are needed to obtain different desired outcomes.

Teacher expertise

In terms of teacher expertise, Shulman (1987) identified several types of knowledge teachers need to possess in order to operate at a high level of competence in their profession. Much of this knowledge is acquired from years of teaching experience and cannot all be instilled into student teachers during initial training. According to Shulman, teachers need:

- Subject knowledge (a deep understanding of their subject matter).
- Pedagogical content knowledge (knowing how to make that subject matter accessible, understandable and interesting).
- General pedagogical knowledge (knowing many instructional methods and management strategies).
- Curriculum knowledge (knowing the content of the syllabus and of any national curriculum).
- Knowledge of learners and their characteristics.

It will be noted that Shulman's categories of knowledge identify several competencies that students themselves expect to find in an effective teacher (see

p. 4); for example, 'knows what he or she is talking about', 'explains clearly', 'controls the class well', 'doesn't give up on you', 'makes allowances'). Having deep subject knowledge, together with good pedagogical content knowledge, are the features that represent teachers who are least likely to confuse students by their teaching or make unreasonable demands. In addition, a good knowledge of students and their learning characteristics is essential if teachers are to attempt to meet their individual needs.

Meeting individual needs

An important component in the effective teaching model is the ability to respond positively and proactively to differences among students. This component is sometimes referred to as 'teaching adaptively' or 'differentiating instruction' (Good & Brophy, 2003; Westwood, 2002). Adaptive teaching could be defined as instruction geared to the characteristics and individual needs of students. Van Kraayenoord and Elkins (1998) have commented that when teaching is carried out as an interactive process, the teacher is sensitive to the characteristics of the students and uses a wider variety of techniques to respond to individual differences. Differentiation can be implemented through the teaching approach, content of the curriculum, assessment methods, classroom organisation, student grouping and teachers' interactions with individual students. Research seems to indicate that, if skilfully implemented, adaptive instruction improves learning (Tomlinson, 2001).

Perhaps the most extreme form of adaptation to differences among students is reflected in moves toward individualised programming. In the past, a great deal has been written about the need to cater for individual differences by providing each student with his or her own unique program and allowing individuals to learn at their preferred rates. At one time this model was almost held up as the 'ideal', with computer-aided instruction (CAI) being one possible way to achieve it. Recent thinking suggests, however, that any extreme form of individual programming is a very difficult approach to sustain, and is not always effective. In contrast, contemporary views on learning place great importance on social interaction among learners — something that disappears if students are working alone.

As Brandt (1992) pointed out, the flaw in programs of individualisation is that the students tend to get further and further apart, not closer in their attainments. Individualisation tends to exaggerate and maintain differences among students instead of closing the achievement gap. Many teachers recognise that the platitude 'a student must be allowed to learn at his or her own rate' is nonsense if the student is actually progressing much more slowly than is necessary. For some students it is essential that teachers intervene to accelerate the rate of learning, otherwise these students simply fall further and further behind. Active and explicit group instruction from the teacher, as described in the next chapter, has the potential advantage of increasing learning rate.

There may be very sound reasons for using individualised programs at times, for example with students of very high ability (gifted students) or for those with severe learning difficulties or disabilities. However, rather than looking toward complicated individualised programming to improve learning outcomes for most students, it is more productive to consider how whole-class teaching, combined with appropriate and flexible grouping and inclusive practices, can be made more adaptive to individual needs. Rather than using individualised instruction, 'differentiation' can be implemented in a manageable way by teaching basically the same curriculum content with all students but tailoring the learning activities and varying the degree of support given to meet the different needs of individual students. Wherever possible, this is regarded as preferable to setting up alternative courses, modifying instructional materials or streaming students by ability. Truly effective whole-class teaching certainly reduces the amount of adaptation needed for differences among students — but some modifications may still be required. This is particularly the case if mixed-ability classes contain some students with disabilities (Janney & Snell, 2004).

Adaptive teaching

Effective teachers recognise the different aptitudes and special learning needs in any group of students, and they plan lessons to accommodate these differences. But teaching adaptively is far from easy. Several studies have indicated that teachers know that they should modify the teaching approach or materials for some students, but find this very difficult to do in practice (Chan et al., 2002; Schumm & Vaughn, 1998). Although in theory there are many potential strategies for modifying methods, curricula and resources, in practice it is not always feasible or desirable to use such strategies within a mixed-ability class. Deschenes, Ebeling and Sprague (1999, p.13) have provided the sound advice that:

> Adaptations are most effective when they are simple, easy to develop and implement, and based on typical assignments and activities. Adapting in this way is feasible for the classroom teacher because it is relatively unobtrusive, requiring little extra time for special planning, materials development, and/or instruction.

Perhaps it is reassuring to find that effective teachers already appear to make several commonsense adaptations to the processes of instruction while lessons are in progress (Scott, Vitale & Masten, 1998). For example, the following tactics are observed during lessons where teachers are sensitive to differences among learners. The teacher:

- Simplifies and restates instructions for some students.
- Reteaches material to certain students, or provides an additional demonstration.
- Gives more descriptive praise to certain students.
- Rewards different students in different ways, and some more frequently than others.
- Sets shorter-term goals for some students.
- Monitors some students more closely than others, and gives more frequent feedback.
- Provides more (or less) assistance to students as they work.
- Accepts different quantities and qualities of book work.
- Asks questions at different degrees of complexity, and targets specific students with questions at an appropriate level.
- Encourages peer assistance.
- Selects or creates alternative resource materials.

Some students need to be taught directly the knowledge and skills that other students acquire easily through incidental learning and informal methods. Failure to adapt a student-centred informal teaching approach may place at risk those students who require direct teaching and a slower pace. This is particularly important in key areas such as literacy and numeracy, where some students appear to benefit most from direct teaching. Equally problematic, failure to extend and challenge gifted children through the provision of more stimulating activities for independent learning may result in boredom and underachievement. Achieving this balance between teacher-directed and student-centred learning is difficult to accomplish. Some of the issues are discussed in following chapters.

It has become clear in recent years that a balanced approach to classroom teaching is required (Ellis, 2005). That balance must be achieved by careful blending of at least three main types of experience (Alexander, 1995):

- Direct teaching, which instructs students explicitly in what to do and how best to do it.
- Inquiry, which poses problems, asks challenging questions, and helps students reflect and think for themselves.
- Scaffolding, which supports students' learning so that they can move forward from their present level of understanding to the next level.

Teacher-directed and student-centred methods that may be used to achieve such a balanced approach are described fully in the next two chapters.

Useful resources

Effective teaching and instructional design

Burden, P.R. & Byrd, D.M. (2003). *Methods for effective teaching* (3rd edn). Boston: Allyn & Bacon.

Freiberg, H.J. & Driscoll, A. (2005). *Universal teaching strategies* (4th edn). Boston: Allyn & Bacon.

These books provide additional information on effective teaching and instructional design.

Differentation

Janney, R. & Snell, M.E. (2004). *Modifying schoolwork* (2nd edn). Baltimore: Brookes.

Westwood, P. (2003). *Commonsense methods for children with special needs* (4th edn). London: Routledge-Falmer.

More information on differentiation can be found in these publications.

Other

Lerner, J. & Kline, F. (2006). *Learning disabilities and related disorders: Characteristics and teaching strategies* (10th edn). Boston: Houghton Mifflin.

Rosenberg, M.S., O'Shea, L. & O'Shea, D. (2006). *Student teacher to master teacher: A practical guide for educating students with special needs*. Upper Saddle River, NJ: Pearson-Merrill-Prentice Hall.

These books are useful for other topics covered in this chapter.

2 Teacher-centred approaches to instruction

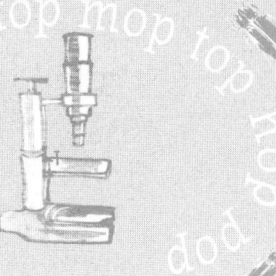

The quality of teaching and learning provision are by far the most salient influences on students' cognitive, affective, social and behavioral outcomes of schooling — regardless of their gender or backgrounds and the schools in which they are enrolled. Indeed, findings from the related local and international evidence-based research indicate that 'what matters most' in 'making schools better' is quality teaching (Rowe, 2004, p.1).

Many attempts have been made to categorise teaching approaches (e.g. Joyce, Weil & Calhoun, 2004; Kauchak & Eggen, 2003). Often, particular approaches or methods are placed somewhere along a continuum from 'teacher-directed instruction' (the *instructivist* perspective) at one extreme to 'student-centred learning' (the *constructivist* perspective) at the other. However, the reality is that most approaches to teaching naturally combine elements of both direct teaching and student-centredness, so when one observes in a classroom it is not easy, in practice, to categorise precisely a particular teacher's overall approach. For example, a lesson may commence with a great deal of teacher direction and explanation, but changes quickly to less structured student-centred activity. During a single lesson in a classroom the instructional techniques may reflect skilful integration of both teacher-directed and student-centred learning methods.

This chapter presents a description of teaching approaches that tend to be located toward the 'teacher-directed' end of the methodology continuum. The main features and assumptions underpinning each approach are discussed, together with advantages and the difficulties or disadvantages associated with its use. In particular, any potential problems or benefits the approach may present for students with learning difficulties are identified. The chapter that follows presents similar information on approaches generally regarded as 'student-centred'.

No attempt is made in this book to suggest that one particular method, when used alone, is superior to all other methods for all purposes. One single method of teaching cannot suit all types of learning, and common sense indicates that different methods are required to achieve different learning objectives. Clearly, methods of teaching should be selected according to their fitness for specific purposes (Galton et al., 1999). An effective educational program usually contains a combination of different methods designed to encourage particular types of learning and to match students' learning characteristics. The selection of an appropriate teaching method depends not only upon the age, ability and developmental level of the students, but also the nature of the subject matter to be taught, the objectives for the lesson, the physical setting in which the lesson will take place, the material and human resources available for use, and the time allocated for teaching the lesson. As Kizlik (2005, p.2) points out:

> There is no shortage of information on what constitutes a particular instructional method. What is far more important is the professional knowledge base that provides criteria for when a particular method is appropriate for given content with students at a defined level of development and who have acquired the prerequisites necessary to learn the content.

This exploration of teaching methods begins by looking at the fairly traditional expository or didactic approach. Many teachers feel comfortable using such an approach, and expository teaching can be effective in achieving certain learning goals.

Expository approach

Expository instruction attempts to present information to learners in a form they can easily access and understand. Expository methods include demonstrating, lecturing, explaining, narrating, requiring students to read a textbook or manual,

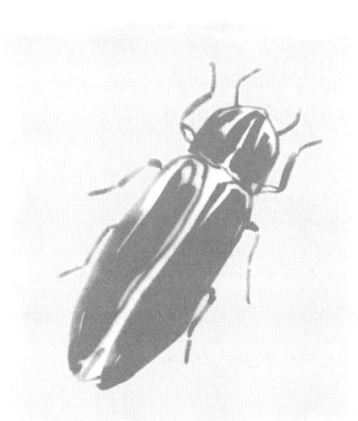

showing students an instructional video, or asking students to work through a computer program presenting information. It is assumed that learners will process new information thoughtfully, and that it will link in an organised manner with their prior knowledge. Ormrod (2000, p.540) remarks, 'Perhaps the major advantage of expository instruction is that it enables students to explore a topic in an organised and relatively time-efficient manner'.

An expository approach accords well with Ausubel's (1968) notion of 'reception learning', because the material to be covered is presented in a fairly complete and meaningful form (McInerney & McInerney, 2002). But the principle underpinning expository instruction is often referred to (in derogatory terms) as 'transmissionist', implying an erroneous belief that knowledge and understanding can be conveyed easily from the teacher to the students by direct verbal and visual means.

Ormrod (2000) reports that teachers can enhance students' learning from expository instruction if they:

- Begin the lesson with an 'advance organiser', clarifying the objectives for the lesson and indicating how the key concepts are interrelated.
- Present points in a clear and coherent sequence as the lesson develops.

- Connect new information with students' prior knowledge.
- Use visual aids to illustrate important points and to hold attention.
- Apply verbal and visual cues to highlight important points; for example, by writing important information on the whiteboard, underlining key words, using bold type or colour to highlight textbook information, and simply stating, 'This is important to remember'.
- Pace the presentation at the optimum rate to allow students time to process information and to take notes if required.
- Summarise key points again to close the lesson, and draw attention again to notes on the whiteboard and to the lesson objectives.

Lecturing — perhaps the most 'teacher-centred' of all methods — is not typical of the approach used in most contemporary primary schools. Even in secondary schools the formal lecture is more likely to be replaced now by 'interactive whole-class teaching' wherein the teacher presents information, asks questions, answers questions from students, and asks students to explain or in some other way to think and respond. (Interactive whole-class teaching is discussed more fully below.) Lecturing is more typically found in universities, where it has a traditional role as a method for delivering course content to large groups of students. The majority of university students would probably say that it is the least effective method of helping them learn subject matter in depth, although the lecture might be an effective way to introduce a new topic that will be investigated and discussed in more practical ways later. Barry (1995) suggests that the lecture strategy is appropriate when the basic purpose is to disseminate information that is not available elsewhere, and to provide an overview of a new topic.

Good and Brophy (2003) believe that lectures, when presented in interesting and enthusiastic ways, can stimulate interest and raise questions that students will want to follow up. Two key factors related to good lectures (and effective whole-class lessons) are *clarity* and the teacher's *enthusiasm*. Sotto (1994) suggests that clarity in teaching is optimum when teachers:

- Know the subject matter well.
- Identify key ideas in what is to be taught.
- Appreciate the subject from the perspective of a novice learning it for the first time.
- Explain things in simple terms.

Eggen and Kauchak (2004) suggest that many of the inherent weaknesses in formal lecturing can be overcome if the presentation time is interspersed with frequent periods of questioning and discussion. This not only makes learners participate more actively, it can also reveal to the lecturer whether the learners understand and relate to the material being presented.

Potential difficulties and disadvantages of expository teaching include the following:

- Expository teaching (particularly lecturing) does not take account of individual differences among students such as prior knowledge, language background, experience or motivation.

- Students with learning difficulties are often at a disadvantage in lessons that rely on expository methods due to their inability to maintain attention in a passive learning situation. Lecturing and explaining place a high premium on learners' ability to concentrate and to follow a line of thought.

- Expository methods also require that learners have adequate linguistic skills, particularly a good vocabulary, and adequate reading and writing skills. Students with learning difficulties often exhibit poor language and literacy skills so they have problems understanding what the teacher says, or they have difficulties processing what is written in the textbook. They also have problems taking notes during the lesson so they are at a disadvantage with follow-up homework assignments and independent study.

- Students with learning difficulties often lack confidence and are therefore unlikely to ask the teacher or lecturer any questions, or to admit that they do not understand something.

Kevin Barry's chapter in the text by Maltby, Gage and Berliner (Barry, 1995) provides a very comprehensive and valuable coverage of the lecture method. Barry discusses how to maximise the effectiveness of lectures. Using a different perspective, Janney and Snell (2004) provide suggestions for adapting oral presentations to suit learners with differing abilities and language competencies.

Interactive whole-class teaching

Interactive whole-class teaching is a way of working with a class of students to ensure very high levels of active participation and a high response rate. The teacher may begin by presenting information or setting a problem using an expository approach, but then students are expected to contribute their own ideas, express their opinions, ask questions, and explain their thinking (Dickinson, 2003; Reynolds & Farrell, 1996). In relation to the learning of basic mathematics, for example, the DfEE in Britain describes interactive whole-class teaching in these terms:

> High quality direct teaching is oral, interactive and lively. It is not achieved by adopting a simplistic formula of 'drill and practice' and lecturing the class, or by expecting pupils to teach themselves from books. It is a two-way process in which pupils are expected to play an active part by answering questions, contributing points to discussions, and explaining and demonstrating their methods to the class (DfEE, 1999, p.11).

One of the advantages claimed for interactive whole-class teaching is that it may help close the learning gap that usually appears between higher achievers and lower achievers when individualised progression ('work at your own pace') methods are used (Zhou, 2001). Studies of teaching methods used in countries where students do extremely well in international achievement tests (e.g. Hungary and Japan) seem to indicate that the teachers employ interactive whole-class teaching methods widely and effectively. As a result of these findings, interactive whole-class teaching has been strongly advocated in Britain for the teaching and improvement of literacy and numeracy skills in primary schools. The approach is seen as more productive than individual programming or unstructured group work for raising standards in basic skills and is certainly regarded as preferable to the boring and repetitive lessons often observed (Harrington, 1999).

Jones and Tanner (2005) have remarked that there are differences among teachers in how they interpret the concept of interactive teaching and how they accommodate it into their own style. It must be noted that recent studies seem to indicate that many teachers do not seem particularly skilled in the use of this interactive approach (Hardman, Smith & Wall, 2003). In many classrooms the approach does not appear to have been implemented effectively. In particular, the quantity and quality of contributions from students during lessons and the quality of feedback they receive from the teacher are not significantly better than occurred with  traditional whole-class teaching (Hargreaves et al., 2003). This may be due to insufficient training of teachers in the use of the interactive method, or perhaps, as Kauchak and Eggen (2003) suggest, some teachers are simply not good at 'thinking on their feet' and responding to students' spontaneous ideas and questions that form a key aspect of the approach.

Other potential difficulties in implementing interactive whole-class teaching include the following:

- Some students do not participate fully in the lesson unless the teacher is very skilled in drawing them in by direct questioning.

- Not all teachers recognise the value of encouraging 'choral responding' (all students answering together) sometimes. Instead, the pace of the lesson is slowed unintentionally by asking individual students to raise a hand if they wish to ask or answer a question.

- On the other hand, if the pace of the lesson is too fast, students with learning difficulties tend to opt out.

- Teachers must not regard interactive whole-class teaching as purely comprising verbal exchanges between teacher and students. Frequent supplementary use must be made of the whiteboard to hold students' attention and to summarise points in the lesson.

For additional information on interactive whole-class teaching, see Hargreaves et al. (2003) and Reynolds and Farrell (1996).

Direct instruction

There are at least two separate but related models of direct instruction. One is the generic form of teacher-directed instruction, as identified, for example, in the 'teacher effectiveness' research of the 1980s and delineated clearly by Rosenshine and Stevens (1986) and Rosenshine and Meister (1995). This generic form has also been termed 'systematic instruction', 'direct teaching', 'active teaching' and 'explicit teaching'. It typically involves clear demonstration, explanation and modelling by the teacher which is followed by guided practice, independent practice and frequent reviews. Direct teaching of this type has proved very effective in raising achievement levels in basic academic skills for all students, and seems to be particularly beneficial for students with learning difficulties. Kindsvatter, Wilen and Ishler (1996, p.231) have remarked:

> As to which learners benefit most from this systematic approach, research tells us that it is helpful for young children, slower learners, and students of all ages and abilities during the first stages of learning informative material, or material difficult to learn.

Although direct teaching takes many shapes and forms, it is particularly associated with a model of instruction made popular in the US by Madeline Hunter (Hunter, 2004). Her approach to lesson planning has been quite influential in many teacher education programs.

The second representation of direct instruction is an even more carefully articulated model, associated in its published form with the pioneering work of Engelmann at the University of Oregon, together with Becker, Carnine, Silbert, Gersten, Dixon and others in the US. This highly teacher-directed form of curriculum delivery uses the capitalised form of title: Direct Instruction (DI).

DI is a rigorously developed, fast-paced method for teaching basic academic skills, providing high levels of productive interaction between students and the teacher. DI procedures are based on behavioural learning principles: clear objectives, modelling, high response rate, reinforcement, error correction, criterion-referenced performance and practice. The approach was originally associated with the commercially produced program DISTAR (*Direct Instructional System for Teaching and Remediation*) covering step-by-step instruction in early reading, language and arithmetic for disadvantaged and at-risk children, but DI now includes materials covering writing, spelling, comprehension, mathematics and problem-solving for a much wider age and ability range (e.g. Dixon & Engelmann, 1976; Engelmann & Bruner, 1995).

The beliefs underpinning DI are that:

- All students can learn if taught correctly.
- Performance expectations should not be lowered for disadvantaged students or those with learning difficulties.
- Low-performing students must be taught more, not less, in the available time if they are to catch up with their peers.
- Basic language, literacy and numeracy skills must be mastered thoroughly to a high degree of automaticity, thus providing a firm foundation for future learning.
- Instruction must be broken down to teachable and learnable steps.

The key features of a DI session are:

- Students are taught in small groups, based on ability.
- The teacher gains and holds students' attention and conducts the lesson.
- Scripted presentation ensures that all steps in the teaching sequence are followed and that all questions and instructions are clear.
- Students actively respond to the teacher's questions or prompts as a group, and individually.
- The teacher gives immediate feedback and correction.
- The lesson is conducted at a brisk pace, with approximately 10 responses from students per minute.

Usually, the children are seated in a semi-circle facing the teacher. The teacher may use a whiteboard, overhead projector or other method to present visual information (e.g. alphabet letters, words, numbers) to the children. Rather than requiring each child to 'raise a hand' to reply, much choral responding by the group is used as a strategy for motivating students and maximising participation. The brisk pace of the session holds children's attention effectively and prevents boredom. For the beginning stages of reading, DI adopts a phonic decoding approach in order to provide students with skills they need when tackling the reading or spelling of an unfamiliar word. In older age groups, and with more advanced work, students have their own print materials or workbooks.

Of all the available methods of teaching in school, DI has probably been the most thoroughly researched (for summaries, see Adams & Engelmann, 1996; Carnine, 2000; Ellis, 2005; Schug, Tarver & Western, 2001; White, 2005). The best-known evaluation was conducted in *Project Follow Through*, one of the most ambitious evaluation projects undertaken in the US. It was a longitudinal study (1967–1976) involving over 70 000 disadvantaged children in the age range K–3. The children were taught either by child-centred constructivist methods or by structured teacher-directed methods. Overall, the results from the study clearly demonstrated the superiority of the direct instructional model for improving achievement in reading, arithmetic, spelling and language, and for boosting children's self-concept and IQ. Carnine (2000) reminds us that only the DI model was effective in getting students to perform at or near the national average in basic maths and literacy. Students in the other treatments (including discovery learning, language experience, developmentally appropriate practice and open education) often performed at a lower level than the control group. Schug, Tarver and Western (2001, p.12) conclude that:

> Direct Instruction has a strong research base confirming its positive effects on student learning. The supporting evidence arises from well-controlled experimental studies that validate the principles and the theory underlying Direct Instruction. In addition, small-scale pilot studies have documented the effectiveness of particular Direct Instruction programs in various classroom settings, and comprehensive evaluations have demonstrated the effectiveness of Direct Instruction more generally across classrooms and schools.

Given this hard evidence, one would expect to find DI being widely used for teaching important academic skills in the early years, but the situation is in fact the very opposite. It seems that teachers strongly prefer to use child-centred

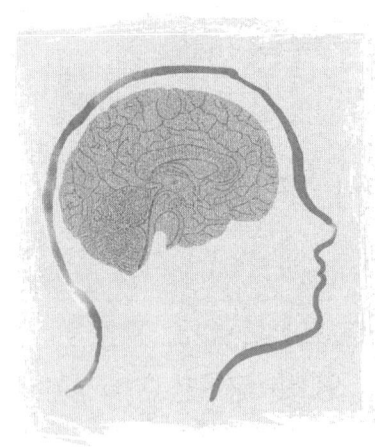

constructivist methods that encourage children to discover and learn at their own rate. From the very start, teacher education institutions tended to ignore DI (unless presented within educational psychology units) so trainee teachers have had little or no opportunity to observe DI lessons in action, and even less chance to develop any skills in using DI strategies.

Although from the late 1970s DI did spread from the US to Britain, New Zealand and Australia, it never became fully accepted as a mainstream method. In Australia, for example, interest in DI first peaked in the early 1980s when several positive evaluation studies were completed (e.g. Lockery & Maggs, 1982), but with the advent of 'progressive' child-centred methods such as 'whole language' and 'process maths', DI fell out of favour. At the present time, however, there are signs of a resurgence of interest in direct teaching in Australia (e.g. Ellis, 2005; Farkota, 2003; Hempenstall, 1996) due in part to a growing dissatisfaction with students' achievements in basic academic skills under predominantly constructivist teaching approaches.

The difficulties and disadvantages associated with the use of DI include the following:

- Many teachers react very negatively toward DI, claiming that it is too prescriptive, too highly structured, too rapidly paced, and places too much focus on basic skills.

- Teachers also claim that DI allows very little opportunity for a teacher or the students to be creative.

- DI must be implemented on a daily basis using small groups rather than a whole class; this can cause problems in scheduling and staffing.

Precision teaching

Precision teaching (PT) is an approach first introduced by the late Ogden Lindsley (1964; 1990; 1992a) and is not so much a teaching method as a system for closely monitoring the effects of any method. It has strong applications within the field of remedial and special school teaching, but can be also applied across the curriculum in areas where skill-building, automaticity or increased work output are required. It has been found that PT can be combined with computer-assisted learning; for example, in using drill and practice software to enhance basic skills in spelling and arithmetic (Harrington et al., 2004).

PT employs a recording method originally used in behavioural research to record changes in response rates, such as increased or decreased speed and fluency in executing a skill. In the classroom context the PT process consists of taking accurate one-minute time samples each day in a selected skill area such as text reading, spelling, writing or carrying out arithmetic calculations (e.g. Updike & Freeze, 2002). Students' responses within the time period are counted, and the results entered on a personal chart. The aim is for students to improve their daily scores until they reach a target level of proficiency where the skill has become automatic. The rate at which a student improves gives the teacher information to help refine the teaching method or to adjust the curriculum materials.

Originally Lindsley devised a special six-cycle semi-logarithmic graph known as the 'Standard Behavior Chart' or 'Standard Celeration Chart' for daily recording of responses. This graph is still used in most clinical teaching situations, but classroom teachers have often modified the chart to suit their own needs. Practical suggestions for teachers on charting students' learning trajectory can be found in Project PRODUCT (n.d.), *Blueprint for PRODUCTive Classrooms*.

The usual PT procedure involves four steps:

1   The teacher first uses test results or work samples to determine a basic academic skill that the student needs to improve. For example, it may be found that the student produces only a few lines of written work in English lessons and needs to increase quantity of output, or the student might be a slow and dysfluent reader, needing to improve speed and accuracy.

2   A specific performance objective is prepared by the teacher, shared with the student, and agreed as a target for the student to achieve. For example, the objective might state: 'The student will read orally a selected passage from two different Level 1 materials on two separate days at 150/250 words per minute, with only 0–2 errors' (Kubina & Starlin, 2003).

3   A series of daily practice sessions are then provided, within which student performance is counted for one minute. Reading rate and error rate are recorded on the student's chart after each practice session.

4   Based on the daily recordings, the teacher makes decisions about the program. For example, the teacher may decide to increase or decrease the level of difficulty of the material the student is using, or may decide to introduce a token reward system to reinforce the student each time the target performance is achieved.

Advantages of using PT include the following:

• Research has consistently reported higher student achievement than is usually obtained by traditional instructional methods in basic skills (Binder & Watkins, 1990).

• Students are also said to gain in self-esteem and to like the recording method because it gives them clear evidence that they are indeed learning (Roach, 2005).

• Intrinsic motivation is enhanced once the student observes his or her own progress and feels successful.

Difficulties associated with using PT include:

• Some teachers are reluctant to use what appears to them to be old-fashioned drill exercises and behavioural methods to achieve the objectives.

• Some teachers are reluctant to impose time-limit pressures on students.

• It is not easy in a large class to find time daily to monitor one student's development so closely.

Given that there is much positive evidence to support the use of precision teaching, it is surprising (and disappointing) that the approach is not more widely used in schools (Lindsley, 1992b). The problem may be due in part to the fact that teachers in training are not exposed to it or encouraged to use it during their teacher education programs.

Mastery learning

Mastery learning (ML) is associated mainly with the work of Carroll (1963), Bloom (1971) and Block (1971). The ML approach gained maximum popularity in the 1970s and 1980s, but is still regarded as useful today. It is not so much a teaching method as a system for designing and delivering an instructional program that enables all students to reach required standards of achievement. Under traditional methods, teachers take students through the curriculum at a common pace and move all students ahead even if some have not yet mastered previous concepts, skills or processes.

This lockstep practice is known to contribute to the creation and accumulation of learning difficulties.

By way of contrast, the core principle of mastery learning is that almost everyone can learn if exposed to good quality teaching and given *sufficient time* for study, practice and application. According to ML principles, it is the time required to learn, not innate ability, that represents the important difference among learners. Whereas traditional teaching methods hold learning time constant and accept wide variation in achievement from student to student, ML instead differentiates the time allocated for learning according to individual need, thus enabling all students to reach the specified standard.

Davis and Sorrell (1995) reviewed the research on effectiveness of ML and found the approach to be generally effective in raising attainment levels and boosting students' feelings of self-efficacy. Cole and Chan (1990, p. 167) remarked:

> In general there is little dispute among educators regarding the positive effects of the mastery learning technique on the intended objectives of instruction. It is also recognised that mastery learning is most appropriate for school subjects that require skill mastery and for students who may not have superior ability.

The basic features of mastery learning include:

- Clear performance objectives to define precisely what a learner must be able to do and to what standard to achieve mastery of the information or skill.

- Content of the course is arranged into manageable teaching and learning units.

- Successful performance in each unit is assessed before a learner progresses to the next unit.

- Instructional materials are selected and teaching takes place.

- The actual teaching method in ML is not prescribed but usually embodies features of direct teaching (modelling, practice, reinforcement, formative assessment, reteaching when necessary and summative assessment).

- Usually a teacher-led, group-based approach is recommended, and students are expected to learn from the teacher and by cooperating with each other.

- Corrective feedback and direction are given to any students who are having difficulties.

- The time a learner will take to reach mastery level will vary from individual to individual according to amount of practice and reteaching required; some of the additional time is represented by private study and homework.

- Computer-assisted learning can be used at both the group instruction stage and at the additional individual practice or correction stage.

Advantages associated with the use of ML include the following:

- Teachers have to plan carefully the scope and sequence of instruction.
- The units are usually task-analysed to identify steps in teaching and to identify potential difficulties.
- Learners are provided with clear and logical tasks and activities.
- Instructional materials are matched to students' abilities and to unit objectives.
- Achievement is judged against the predetermined objectives.
- Students receive frequent feedback and correction.
- Remedial teaching or extra coaching is provided if a student is not reaching mastery.
- ML can break into the failure cycle by helping lower ability students experience more success.
- The ML design has much to offer in traditional remedial or resource-room tutorial programs.
- Research evaluating the effectiveness of ML has consistently proved it to be helpful in raising standards and in building students' confidence (Huitt, 1996).

Disadvantages and difficulties in using ML include the following:

- Teachers have difficulty planning and preparing materials and formative tests for ML courses (very time-consuming, and requiring sound curriculum design skills).
- In practical terms it is not easy to allow students to work at different rates and to move from unit to unit at different intervals; grouping, testing and scheduling become difficult.
- Fast learners may be frustrated if they have to wait while slower students spend additional time reaching minimum standards (although it is suggested that fast learners can proceed to extension or enrichment activities).
- ML usually requires the adoption of a whole-school approach if it is to be adequately supported; it is very difficult to implement and sustain in a single classroom.
- The logistical problems of implementing ML successfully cannot be ignored. Good and Brophy (2003) report that in practice most teachers using the approach do not manage to allow students as much time as they really require to reach mastery, but instead move all students ahead after two attempts to reach a criterion, whether successful or not.

Computer-based learning

Computer-based learning (CBL) could just as appropriately be placed in the following chapter on student-centred constructivist approaches. However, since the computer often acts in the role of teacher in many of its educational applications, it has been included here. In the following chapters frequent mention is made of the contribution that information and communication technology (ICT) can make within a constructivist approach — for example, when students are independently searching for information to complete a project or to help solve a problem.

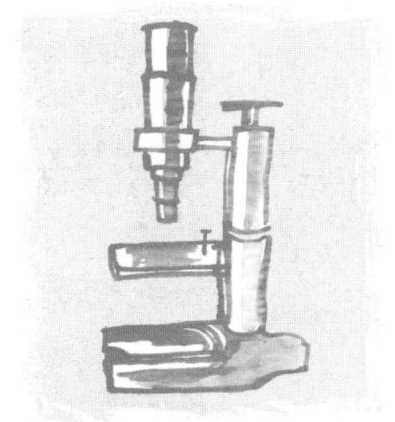

Computers in the classroom have provided learners and their teachers with additional fast and simple ways of accessing information, communicating electronically with others and producing high quality written work and graphics. They can also deliver instructional programs covering virtually any area of the curriculum and geared to any age or ability level. Computers and their associated software present great opportunities for motivating students and for improving the quality of educational programs. The use of ICT continues to grow rapidly in the schools of most developed countries, and increasing numbers of students also have access to a personal computer at home. It is probably true to say that many of these students are more 'computer literate' than some of their teachers.

Several terms tend to be used, almost interchangeably, to identify the teaching roles of computers and ICT:

• The most general term is computer-based learning (CBL) or computer-based instruction (CBI). This term implies that computers represent the principal teaching and learning resource used in the educational environment, and that the curriculum is implemented largely through the use of information and communication technology. Interactive learning programs are typically loaded onto a student's computer in the form of CD-ROM or via an Internet connection. Instructional programs are often presented in a navigable form, where students can link with other online resources when appropriate. Students may work individually at the computer, with a partner, or occasionally as a small group according to the nature of the program.

• *Computer-assisted learning or computer-aided learning* (CAL) is the term favoured by educators who regard the use of ICT as an adjunct to the general teaching approach used in the classroom program. Much of the learning in that program will be achieved through other forms of presentation and study, with ICT being used only to support learning as necessary. An example might be using drill and practice software to help certain students increase automaticity in number skills or spelling. All students can use ICT as an aid to conduct projects and investigations, search and find information (including graphics), produce original printed work of high quality, send email to communicate with others, or contribute personal ideas online to a classroom 'learning space' or 'chat room' (Kennedy, 2000). Such lessons may occur regularly throughout the school day, or may be confined to specific periods in the week.

- *Computer-supported collaborative learning* is a term which covers a range of models, from two or more school students working together on a software package or using ICT to help research a topic, through to quite complex arrangements in tertiary education for linking distance-education (external) students with each other and with their tutor (Ewing & Miller, 2002).

Computer software for educational purposes includes:

- *Drill and practice programs:* Used in tutorial or remedial contexts for building students' skills in areas such as phonics, reading, arithmetic or spelling.

- *Interactive tutorial programs:* These introduce new information or skills. The content may be presented through pictures, video clips, on-screen text, sound commentary and/or accompanying supplementary print materials. The student is required to make active responses at frequent intervals to demonstrate understanding or application. The program immediately provides corrective feedback. The materials may also have an element of additional practice contained within them to consolidate learning.

- *Simulations:* These include role-plays and other visual representations of real-life situations, or demonstrations of steps in a process or procedure. Often used as supplements to tutorial-type material, or to set up scenarios for exploration or problem-solving (see 'Situated learning' and 'Anchored instruction' in the next chapter).

- *Problem-solving programs:* May contain simulations, or may present data in some form (e.g. spreadsheet). Students must work with the information to solve a problem and make decisions. They may also need to do additional Internet searches and navigate to other online resources.

- *Computer games:* These may be mainly recreational, but many games can also stimulate a student's logical thinking, planning and decision-making. The games format is also effective in holding students' attention.

In general, the findings from research into the effectiveness of CBL have been positive (Cole & Chan, 1990; Linden, Banerjee & Duflo, 2003; McInerney & McInerney, 2002). Some studies suggest, however, that CBL does not necessarily produce significantly higher achievement in students than would occur under good quality teaching by more conventional methods. As Ormrod (2000, p.553) wisely comments:

> A computer can help our students achieve at higher levels only when it provides instruction that we cannot offer as easily or effectively by other means. There is little to be gained when a student is merely reading information on a computer screen instead of reading it in a textbook.

Good and Brophy (1997, p.319) have summarised the overall impact of CBL thus:

> Effects are stronger with programs that involve tutorials, rather than simple drill and practice, with younger rather than older students, and with lower-ability or remedial students rather than other students.

The appropriate use of CBL can certainly have a positive influence on students' enthusiasm, motivation and concentration — and these are important considerations when working with students who have learning difficulties or disabilities. Polloway and Patton (1997, p.134) state that:

> Students with disabilities frequently express positive attitudes toward computers. They are eager to participate in computer activities and exhibit high levels of on-task behaviour and improved instructional-related performance and behaviour.

In the remedial teaching field, CAL has become increasingly valued as a method not only for building automaticity in basic skills but also improving confidence, motivation and attention to task (Kennedy, 2000). Meta-analyses of results from remedial intervention studies fairly consistently find CAL to be moderately effective (Kavale & Forness, 2000; Swanson, 2000b). For example, the use of relevant learning software has proved helpful in working with dyslexic students (Minton, 2002; Prideaux, Marsh & Caplygin, 2005), dysgraphic students (DO-IT, 2002) and in early intervention literacy programs (Kingham & Blackmore, 2003; Polkinghorne, 2004). CAL has also been used effectively in teaching basic academic skills to adults with learning problems (Cromley, 2005).

Advantages of CBL and CAL include the following:

- Students make active responses and are 'in charge' of the learning situation.
- Immediate reinforcements and corrective feedback are provided in most tutorial-type programs.
- Learning can be achieved at an appropriate pace for the student.
- Working at a computer is motivating, challenging but non-threatening.
- Students are helped to move toward greater independence in learning.
- CAL is a private method of responding, and students can self-correct mistakes.
- ICT facilitates the production of high quality, well-presented documents.
- Students can be provided with extra practice and overlearning to master basic skills.
- Most students enjoy working at the computer more than using textbooks and print resources.
- Software can be matched to a student's ability level and is therefore one way of individualising learning.
- The teaching of science, social studies, mathematics, environmental education and the arts can be enhanced by documentary or simulation programs and by giving access to Internet resources.

- Programs can stimulate inductive learning through presenting complex and interactive problems.

Disadvantages and difficulties associated with CBL include the following:

- Some teachers lack confidence or expertise in integrating CBL into the curriculum.
- There may be a shortage of computers in the school, or computers may only be available in the computer lab at limited times each week.
- Technical failures occur.
- There are additional demands on teachers' planning and preparation time.
- A few students do not like to learn by ICT methods and prefer group interactions with peers and the teacher.
- Some published software, supposedly for educational purposes, turns out to be entertaining but low in educational value.
- Students with literacy problems have difficulty comprehending verbal information on the screen.

Multisensory teaching methods

Multisensory teaching simply means involving as many senses as possible in the learning process in order to enhance awareness, attention and memory. Multisensory teaching methods are typically associated with intensive individual remedial teaching for students with learning disabilities. Indeed, the best-known multisensory methods (Fernald VAK Method, the Orton-Gillingham Approach and the Slingerland Approach) were developed from work with dyslexic students (Heinz, 2000; Henry, 1998). A multisensory approach is also advocated for teaching students with moderate to severe intellectual disability and has its roots in sensory training approaches to retardation recommended by very early pioneers in the field such as Itard (1962), Seguin (1866) and Montessori (1919).

Multisensory methods deliberately involve the learner in simultaneous use of visual, auditory and kinaesthetic modalities (VAK). The approach is used, for example, when attempting to learn and remember letters and the sounds they represent (phonics), or sight words and numerals. When learning a new word the student will be instructed to look intently at it, hear the word pronounced by the teacher or tutor, pronounce the word in imitation of the model, trace a finger over the outline of the word, trace the word in the air, and then write the word several times while saying the word aloud. In some programs, the teaching materials are textured so that when the learner handles them or traces over them with a finger a tactile sensation is also added, thus making the approach visual, auditory, kinaesthetic and tactile (VAKT).

On the issue of the effectiveness of VAKT methods, the International Dyslexia Association (IDA, 2000, p.1) has stated:

There is a growing body of evidence supporting multisensory teaching. Current research, much of it supported by the National Institute of Child Health and Human Development converges on the efficacy of explicit structured language teaching for children with dyslexia. Young children in structured, sequential, multisensory intervention programs, who were also trained in phonemic awareness, made significant gains in decoding skills. These multisensory approaches used direct, explicit teaching of letter-sound relationships, syllable patterns, and meaning word parts. Studies in clinical settings showed similar results for a wide range of ages and abilities.

Perhaps multisensory approaches that use several channels of input help a student to integrate and store in long-term memory what is seen and heard, whether it be a letter or a word. But VAKT approaches may actually succeed where other methods fail because they cause the learner to focus attention more intently on the learning task. Whatever the underlying reason, this teaching approach, which brings vision, hearing, articulation and movement into play, does appear to result in improved assimilation and retention of taught material.

In a much broader sense, multisensory learning is also involved in most computer-based and multimedia learning programs where visual images are accompanied by sound commentary, and students look, listen, and respond actively with a keyboard or clicks of the mouse. Staley (2005) regards this multisensory input and student involvement as one of the great strengths of computer-based learning and ICT.

Difficulties and disadvantages in using multisensory methods include the following:

- The method is labour-intensive in the sense that it can only be conducted one-to-one.
- Mastery of a single step in the process may take a considerable time.
- While the approach is feasible in clinical and remedial teaching, it is difficult to implement with one student in the context of a large class.

In the following chapter, attention will be given to methods that are regarded as more student-centred. These methods are thought to be entirely compatible with constructivist views of learning.

Useful resources

Direct Instruction

Carnine, D. (2000) *Why education experts resist effective practices (and what it would take to make education more like medicine)*. Washington, DC: Thomas B. Fordham Foundation. Viewed 1 February 2006, (http://www.edexcellence.net/doc/carnine.pdf).
More information on the general reluctance of schools to adopt Direct Instruction can be found in this paper.

Kozloff, M. (2003) *Main features of Direct Instruction*. Wilmington, NC: University of North Carolina. Viewed 8 February 2006, (http://people. uncw.edu/kozloffm/difeatures.doc).
This article provides an excellent overview and examples of Direct Instruction.

Watkins, C. & Slocum, T. (2004). The components of Direct Instruction. In N.E. Marchand-Martella, T.A. Slocum & R.C. Martella (eds) *Introduction to Direct Instruction* (pp.28–65). Boston, MA: Allyn & Bacon.
For more information on Direct Instruction.

Precision teaching

Various authors (2003). *European Journal of Behaviour Analysis*, 4, 1.
A special issue on the topic of precision teaching.

Mastery learning

Cole, P. & Chan, L. (1990). *Methods and strategies for special education*. New York: Prentice Hall.
Provides an excellent coverage of mastery learning and its potential value for students with learning difficulties.

Gentile, J.R. & Lalley, J.P. (2003). *Standards and mastery learning*. Thousand Oaks, CA: Corwin.
This text is a valuable source of additional information on mastery learning.

Computer-based instruction and computer-based learning

McInerney, D.M. & McInerney, V. (2002). *Educational psychology: Constructing learning* (3rd edn). Sydney: Prentice Hall.
This text provides a valuable overview of the various applications of ICT in schools and also discusses the relation of CBL to behavioural and constructivist theories of learning.

Norman, K.L. (1997). *Teaching in the switched on classroom: An introduction to electronic education and HyperCourseware*. College Park, MD: University of Maryland. Viewed 31 January 2006, (http://www.lap.umd.edu/SOC/ sochome.html).
For additional information on computer-based instruction.

Multisensory teaching

Birsh, J.R. (2005). *Multisensory teaching of basic language skills* (2nd edn). Baltimore: Brookes.
Contains useful additional information on multisensory teaching.

3 Student-centred approaches to learning

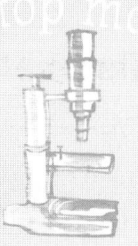

> . . . cognitivists believe that students need to *understand* in order to learn. Students need to participate in how they are being taught, and to reflect on what they have done, in order for learning to take place. The teacher is seen as a *facilitator* of learning, rather than an instructor, thus allowing students to discover for themselves and providing them with time to reflect (Ellis, 2005, p.20).

As Ellis (2005) indicates in the quotation above, student-centred approaches reduce the instructional role of the teacher and place more responsibility on students to engage in and regulate their own learning. Teaching methods that are loosely described as 'student-centred' tend to be aligned with 'constructivist' (or 'cognitive') theories of learning — although some of these methods were in operation long before constructivism emerged as a coherent theory of human learning. The principles underpinning student-centred approaches can be traced back in education theory to advocates of active and experiential methods, such as Dewey (1933), Rousseau (1979) and Piaget (1971).

It will become clear as this chapter develops that various constructive approaches have been given specific titles by their creators, but the principles and practices associated with many of the approaches are very similar indeed. For example, 'situated learning', 'anchored instruction' and 'task-based learning' share many features in common; and the approach known as 'problem-based learning' encompasses all these and several other methods described here. The underlying principles for most of the methods are that:

- The students should be actively involved in the learning process and intrinsically motivated.
- As far as possible, topics or subject matter being studied should be authentic, interesting and relevant.

The subtle differences among the methods described here are usually associated with the varying amounts of guidance and structure provided by the teacher during the learning process, and with the degree of autonomy demanded of the learners.

Discovery learning

Discovery learning (DL) is an inquiry-based approach wherein students develop knowledge related to a topic largely through their own endeavours, using whatever human and material resources they may need. The emphasis is on students being active investigators rather than passive recipients of information delivered to them by the teacher or textbook. Discovery learning is associated particularly with the 'active learning' principles of Dewey (1933) and Bruner (1966). The underlying constructivist principles of discovery learning also accord strongly with Piaget's (1963; 1971) views on cognitive development.

Ormrod (2000, p.541) defines discovery learning as:

> . . . a process through which students interact with their physical or social environment — for example, by exploring and manipulating objects, performing experiments, or wrestling with questions and controversies — and derive information for themselves.

There are two main forms of discovery learning — 'pure discovery' and 'guided discovery'. In activities that involve pure (unstructured) discovery, learners are given little or no direction from the teacher (although any necessary 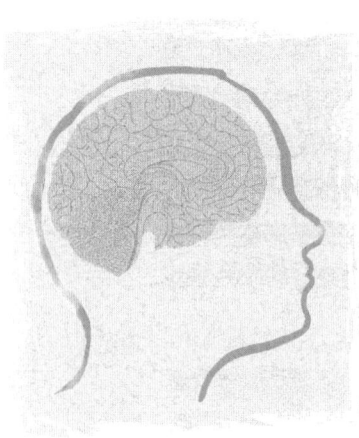 resources may be provided for them). They must decide for themselves the most appropriate method for investigating the given problem and must then reach their own conclusions from their data and the observations they make. This unstructured approach is sometimes used in science, mathematics and for topics in social studies, but the outcomes are not always good, particularly for students with poor study skills, weak self-management and difficulties with inductive reasoning. Eggen and Kauchak (2004, p.497) reviewed research on the effectiveness of discovery learning and conclude that 'Students in unstructured discovery activities become lost and frustrated; this confusion sometimes leads to misconceptions'. Often students with learning difficulties do not have a clear idea of what they are expected to do, and because of weak self-efficacy they do not believe in their own ability to understand a problem by thinking in an active way.

Guided discovery, on the other hand, has a much tighter structure. The teacher usually sets clear objectives, provides initial input or explanation to help students begin a task efficiently, and may offer suggestions for a step-by-step procedure to find out the target information or to solve the problem. Teachers have found that discovery learning is more successful when the discovery process is carefully structured and students have the prerequisite knowledge (Conway, 1997; Tuovinen & Sweller, 1999). In general, guided discovery methods, particularly in subjects such as science, can produce good results (Eggen & Kauchak, 2004).

In order to participate successfully in discovery activities, learners must have adequate inductive reasoning ability. Inductive reasoning is involved in formulating general principles or rules based on experiences with examples and non-examples. In typical discovery learning situations in mathematics or science, examples and non-examples of specific concepts are available to learners, and from these examples they must 'discover' the corresponding rule or relationship. It is also important that students can cope with an element of 'cognitive conflict' when faced with new information that is at variance with their current beliefs and understandings. There is some evidence that higher achieving students eventually benefit from some degree of cognitive conflict and ambiguity, but lower achievers are confused by it and benefit more from direct teaching that minimises ambiguity (Zohar & Aharon-Kravetsky, 2005).

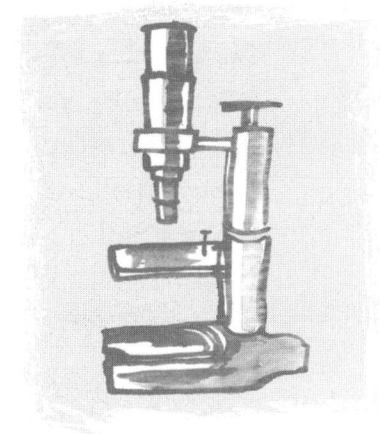

A typical guided discovery learning session may take the following format:

1   An issue or problem is identified.

2   Teacher and students work together to brainstorm ways of investigating the problem.

3   Teacher and students together agree on what forms of data need to be obtained and potential sources of such data (usually the teacher will have gathered such resources already so they are available when needed).

4   Students work individually or in small groups to obtain, analyse and interpret data.

5   Inferences and tentative conclusions are drawn, shared across groups and modified if necessary.

6   Teacher summarises and consolidates the conclusions or solution.

Advantages of DL are considered to be the following:

• Learners are actively involved in the process of learning, and the topics are often intrinsically motivating.

• The activities used in discovery contexts are usually more meaningful than the typical classroom exercises and textbook study.

• Learners acquire investigative and reflective skills that could be generalised and applied in many other contexts.

• The approach builds on learners' prior knowledge and experience.

• It encourages independence in learning.

• Learners are more likely to remember concepts and information if they discover them on their own.

• It fosters group-working skills for collaboration.

McInerney and McInerney (2002) suggest that discovery activities help students learn how to discover, 'how to learn', and how to organise what they have learned. When implemented well, DL is a valuable method for establishing processes of inquiry and encouraging students to engage in reflection and critical thinking. For students with learning difficulties there is some evidence that carefully guided discovery learning results in longer term retention of information and a greater likelihood that new knowledge or skills will generalise (Bay et al., 1992).

Disadvantages of DL include the following:

- DL can be a very time-consuming method, often taking much longer for information to be acquired than would occur with direct teaching.
- DL usually requires a resource-rich learning environment.
- Effective DL usually depends upon learners having adequate literacy, numeracy and independent study skills.
- Students may learn little of value from discovery activities if they lack an adequate knowledge base for interpreting their discoveries accurately.
- Although learners become actively involved, they may still not understand or recognise the underlying concept, rule or principle: in other words, 'activity' does not necessarily equate with 'learning'.
- Young children (and older children with learning problems) often find it difficult to form opinions, make predictions or draw conclusions based on pieces of evidence. Teachers sometimes overestimate children's ability to think as adults.
- Teachers are not necessarily good at creating and managing discovery learning environments, resulting sometimes in poor outcomes.
- Teachers may not monitor activities effectively, so are not able to give the individual encouragement and guidance (scaffolding) that is frequently needed by learners.

It can be concluded that discovery learning can be a valuable approach for achieving certain learning outcomes concerned with process rather than product. Even Bruner (1966) recognised that first-hand discovery is not appropriate or necessary for bringing about all types of learning. Yates (1988), for example, highlighted the inappropriate use of discovery in relation to acquisition of basic academic skills. He wrote that 'requiring a child to actively discover his or her way toward a basic knowledge of numeracy and literacy is to confront the child with tasks of immense difficulty' (Yates, 1988, p.8).

Resource-based learning

The approach referred to as resource-based learning (RBL) can be considered another form of inquiry method. It is closely associated with problem-based or issues-based learning, and is underpinned by constructivist learning principles.

Although there are many different definitions of RBL, the approach is usually described as a methodology that allows students to learn from their own active and creative processing of information using a range of authentic resources (Australian School Library Association and Australian Library and Information Association, 2001). RBL is suited to most areas of the school curriculum and is said to be adaptable to students' different styles and abilities. One of its primary goals is to foster students' autonomy in learning by providing opportunities for them to work individually or collaboratively while applying relevant literacy, numeracy and study skills to investigate interesting topics. As a motivating approach, RBL can be introduced in primary schools, but its chief contribution at the moment is probably in secondary schools and in higher education.

Typically, in RBL situations the teacher introduces an issue, topic or problem to be investigated through the use of relevant resources, including books, magazines, newspapers, DVDs, CD-ROMs, the Internet and human resources (individuals with relevant knowledge and expertise). The teacher and students clarify the nature of the task together and set the goals for inquiry; students then work individually or in groups to carry out the necessary investigation over a series of lessons. RBL is intended to develop and extend six key competencies, namely:

- *Defining:* This is the initial step toward understanding the topic and setting the parameters. Teacher guidance at this stage may involve brainstorming for ideas, concept mapping and providing advance organisers. The teacher helps students prepare for the task.

- *Locating:* Students use library catalogues, electronic searches, CD-ROMs, telephone calls, interviews, emails and letter-writing to begin to identify sources of the information they may require. They also use materials that the teacher has already collected in preparation for this work.

- *Selecting:* From the various sources of information the students extract and interpret relevant data, take notes, reject irrelevant material and summarise key points.

- *Organising:* Students integrate information, review, edit and decide on presentation formats.

- *Presenting:* Students, in conjunction with the teacher, decide on the best means of demonstrating their new knowledge. This could be via essays, reports, dot-point summaries, notes, posters, performances, PowerPoint presentations or oral reports. Presentations can be individual or group-based.

- *Evaluating:* This involves self-assessment, but may also include peer and teacher feedback.

For some students, these six key competencies may not be fully established at the time RBL is introduced, and Cartwright (2001) recommends that explicit preteaching be provided in areas such as locating information, extracting relevant data, summarising and taking notes.

The advantages claimed for RBL include its potential to:

- Motivate all students.
- Stimulate the development of thinking (problem-solving, reasoning and critical evaluation).
- Develop a deeper understanding of subject matter.
- Encourage self-directed learning and reflection.
- Strengthen the independent use of research and study skills using print and electronic media.
- Facilitate an increase in ICT competence and confidence.
- Increase the likelihood that study skills will generalise to other areas of the curriculum.
- Foster enthusiasm for learning through active participation and successful outcomes.
- Increase academic engagement time, compared to traditional teaching methods.
- Improve attitudes towards reading for information, the library and the resource room.
- Allow the teacher to circulate more freely during lessons to help or support individual students when required.

The potential difficulties associated with RBL are similar to those identified for discovery learning, namely that:

- Ideally, RBL requires a resource-rich learning environment, including easy access to computers.
- Effective engagement in RBL depends upon students having adequate literacy, numeracy and independent study skills.

- RBL demands reasonable self-management from the students.
- Some students will learn little from RBL if they lack the prerequisite knowledge for interpreting new information (although they may look busy during the lesson).
- Teachers are not necessarily good at identifying suitable topics for RBL and creating the necessary resources, resulting sometimes in poor outcomes.
- Teachers may not monitor activities effectively, so are not able to give the encouragement and support that is frequently needed by learners.

Project-based learning

In a project-based approach students work individually, with a partner or in groups to gather information on a particular topic or to investigate a real-world issue. They then share with others what they have learned. The extended time frame often provided for project work allows students to plan carefully, revise and reflect upon their learning (San Mateo County Office of Education, 2001). One of the advantages of integrating information technology into project work is that students can learn both ICT skills and specific content knowledge simultaneously (OTEC, 2005).

Project-based learning in various forms has been popular for very many years and represents another approach to student-centred learning based on socio-constructivist principles. The Buck Institute for Education (2002, p.2) defines project-based learning as:

> A teaching method that engages students in learning knowledge and skills through an extended inquiry process structured around complex, authentic questions and carefully designed products and tasks. This definition encompasses a spectrum ranging from brief projects of one to two weeks based on a single subject in one classroom to yearlong, interdisciplinary projects that involve community participation and adults outside the school.

Project-based learning is described as 'interdisciplinary' in the sense that while undertaking a project students may need to draw on information, skills and strategies from mathematics, language and literacy, geography, science, graphic arts and technology (Solomon, 2003). In many ways, project-based learning is similar to resource-based learning because students use books, community publications, reports, online material and other resources to obtain their information; they must then analyse and critique it before organising it in an appropriate form for presentation.

The advantages of project work are considered to be many. According to Kraft (2005) and Buck Institute for Education (2002), project-based learning has the following benefits when compared with traditional textbook teaching:

- It utilises a hands-on approach and 'real-world' orientation.
- The learning process involved in gathering data is valued as well as the product.
- Projects promote meaningful learning, connecting new learning to students' past experience and prior knowledge.
- Students are responsible for their own learning, thus increasing self-direction and motivation.
- Students have ownership of their learning.
- The method utilises various modes of communication and representation.
- It encourages decision-making, and allows for student choice.

- It is an inclusive approach, in that all learners can participate to the best of their ability.
- It encourages use of higher order thinking and conceptual learning, as well as acquisition of facts.
- It develops deeper knowledge of subject matter.
- Assessment is performance-based.
- Self-assessment of learning is encouraged.
- It increases team-working and cooperative learning skills.
- In some situations it is superior to traditional teacher-directed lessons.

Potential difficulties in project-based learning include the following:

- Some students lack adequate baseline skills for researching and collating information.
- When working on projects, some students may give the impression of productive involvement but may in fact be learning and contributing very little.
- When projects involve the production of posters, models, charts, recordings, photographs and written reports on display, there is a danger that these are actually 'window-dressing' that hides a fairly shallow investigation and understanding of the topic.
- When different aspects of a topic are given to different group members to research, there is a danger that individual members never really gain an overall understanding of the whole topic.

Problem-based learning

Problem-based learning (PBL) — also known as 'issues-based learning' — has gained popularity in recent years as a method for use in higher education, particularly in the medical, therapeutic and other professional fields where the 'problem' is often in the form of a 'case study'. Lee (2001, p.10) has suggested that: 'Learning through problem-solving may be much more effective than traditional didactic methods of learning in creating in the student's mind a body of knowledge that is useful in the future'.

Problem-based learning has also been introduced into some primary and secondary schools in curriculum areas such as geography and environmental education where it might, for example, focus on a problem of conservation of vegetation in the local district. PBL has also carved out a place within programs for gifted students (Coleman, 2001). King (2001, p.3) states:

> PBL offers a mode of learning which might be considered closer to real life. This real-life link is twofold: firstly, the projects or problems used often reflect or are based on real-life scenarios; secondly, the processes of team-working, research, data collection, critical thinking and so on are those which will be of use to the students in their further careers.

In problem-based learning, students are presented with a real-life situation or issue that requires a solution or a decision leading to some form of action. With older learners, the problems are often intentionally 'messy' (ill-defined) in the sense that not all of the information required for solution is provided in the problem, and there is no clear path or procedure to follow (Kauchak & Eggen, 2003).

An example of a problem might be that in recent months the density of traffic is increasing in the streets near the school, and the single marked crossing by the school gate is proving inadequate to ensure safety of students and other pedestrians: what can be done? Students work in small collaborative groups and the teacher or tutor is available as a resource. The teacher has the role of general facilitator, not director, of the group activities. Students take responsibility for the group discussions, decisions and actions. This whole process involves all of the usual cognitive strategies required in typical problem-solving (Corrie, 1995; Stanford University, 2001) and may take more than a week to complete. The process involves the following:

1   First, students must clearly identify relevant features of the problem, perhaps representing it diagrammatically or numerically.

2   They must identify any additional information and resources they will need.

3   Next they must generate ideas for ways of tackling the problem and finding a solution. Then they must decide on a line of action.

4   At this point they may establish some short-term goals and allocate specific tasks to different team members.

5   Working with all the information, they then develop a possible solution and make recommendations.

6   If feasible, their recommendations are implemented and the outcomes evaluated.

As Esch (1998) points out, problem-based learning assignments vary widely in scope and sophistication. For younger children, or for students with learning difficulties, it is best to introduce the approach gently at first through well-defined problems that can be solved simply by drawing on the participants' prior knowledge. Later it may be possible to move the children on to more challenging and open-ended topics. Esch (1998) points out that some problem-based approaches place importance on the final outcome and require students to define the problem clearly and arrive at a feasible solution. Other PBL approaches focus much more on the value of the learning and information-gathering process itself, placing less emphasis on quality of any final solution.

The advantages of PBL are considered to be as follows:

- Authentic objectives link school learning with the real world.
- There is greater motivation for learners.
- Students' active involvement leads to effective construction of new knowledge.
- Information and skills from different disciplines are integrated.
- Learning achieved is likely to be retained, and can be transferred to other situations.
- It encourages self-direction in learning.
- It prepares students to think critically and analytically.
- The communication skills and social skills necessary for cooperation and teamwork are enhanced.
- It empowers students to identify, locate and use appropriate resources.

Some of the difficulties associated with using PBL in the classroom are to some extent the same difficulties associated with engaging in any form of problem-solving. Weiten (2001) has identified these as:

- An inability to identify and separate out irrelevant information from what is relevant for addressing the problem.
- A lack of flexibility that causes the individual to think too narrowly of ways in which specific objects or persons can be used.
- A mental set that causes a person not to be able to change the way he or she is thinking about the problem.
- A tendency for the learner to place unnecessary constraints on the way a solution could be implemented (e.g. 'Oh! I didn't realise that I could change that variable!').

Eggen and Kauchak (2004) and Ormrod (2000) have identified additional factors associated with difficulties in engaging successfully in PBL, including:

- Lack of experience in working through problems without direction.
- Lack of specific subject knowledge about the topic or content of the problem.
- A tendency to decide on a solution too early in the process and then resist a change later.
- Limitations in working memory capacity.
- Not all students like group work or discussion, preferring instead to work independently.
- Groups don't always work effectively without support and some direction. Even at tertiary education level, students are often not capable at first of executing the tasks associated with PBL independently, and require direction (Stanford University, 2001).

Task-based learning

The approach known as task-based learning (TBL) is most frequently associated with the teaching of a second language, where it is often referred to as 'task-based language learning' (TBLL). TBLL is believed to be an effective way of developing students' communicative competence. The approach can, however, be used more widely across the curriculum and it shares principles in common with 'resource-based learning', 'problem-based learning', 'project approach' and 'situated learning' (see pp. 45–46). The 'task' set in typical TBL contexts across the curriculum is often in the form of an assignment requiring students to investigate and report on a contemporary issue, to solve a problem or to collect and collate information around a central theme or topic in a specific subject area. TBL is used in primary and secondary schools, but also has applications in higher education (Garner, 2005).

The traditional approach for language teaching has been described as 'presentation, practice and production' (PPP). The teacher first presents or models the language principle for the day, then sets students oral and written exercises (drills) to practise the concept. Finally the students demonstrate their understanding by producing something original (oral or written) that embodies the language principle. In recent years, teachers have come to regard this approach as artificial, because the topics for study are predetermined by the teacher and the textbook and do not arise out of any genuine need the students may have to communicate. Using real situations to stimulate language is one obvious way of overcoming this objection.

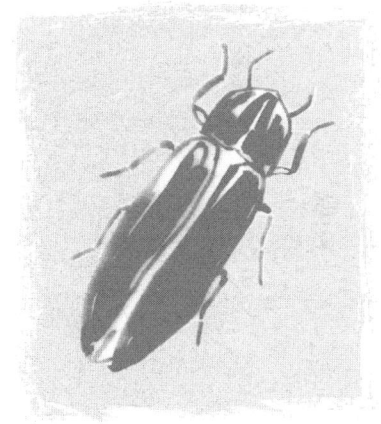

When applied in language teaching, TBLL uses authentic problems or contexts that require verbal interaction among learners to achieve a solution or to reach a goal. Language development is stimulated and enhanced through the communication requirements of the task, and natural opportunities emerge for discussion, questioning, clarifying and decision-making in relation to opinions, actions or options. Learners can stretch their existing skills by using language in original and functional ways. It is believed that through TBLL they will develop a much greater autonomy and confidence in communication than is usually possible using the PPP approach and textbooks. People learn language most effectively in the context of using it for ordinary communication (Brown, Collins & Duguid, 1989) and learning that takes place in a 'real' situation is unlikely to have the problems of transfer and generalisation that occur with decontextualised learning episodes (Chai & Hannafin, 1995). Similar principles underpin a new teaching approach termed 'content and language integrated learning' (CLIL), designed for second-language learners (Clegg, 2005). In CLIL, topics from subjects such as geography or science become the vehicle through which language skills are developed, at the same time as important everyday concepts and understandings are acquired.

Main interest in TBL in language teaching seems to have started with the book *A framework for task-based learning* by Jane Willis (1996). Willis defines a 'task' as a goal-oriented activity with a clear purpose. To complete a communication task requires learners to achieve an outcome and to create a final product that can be shared by others. The model of TBLL proposed by Willis is based on three stages:

1 *Pre-task stage:* The teacher introduces the task (topic) and discusses possibilities with the students. Useful words and phrases are suggested that may be required when talking about and carrying out the task. Learners may be given something to read that is directly related to the task before commencing work.

2 *Task cycle:* The learners then work on the task in pairs or small groups, giving opinions, asking and answering questions, commenting on the ideas of others and asking for clarification when necessary. The students write notes or prepare a report describing how they completed the task and the conclusions they reached. They present their findings to the class in spoken or written form, or groups exchange and read their written reports. Sometimes oral reports or group discussions are recorded for later analysis or study.

3 *Language focus stage:* The learners, under the guidance of the teacher, analyse and reflect upon specific language features evident in the group discussions or reports. The teacher gives feedback and, where necessary, provides students with additional practice in any new language forms or vocabulary that could be linked with the task.

There is no limit to the variety of tasks that can be used for TBLL: for example, puzzles, word problems, building a model, creating a role-play, attempting a reading comprehension exercise or completing a small project. Perhaps an ideal task is one that can be completed to different acceptable standards according to students' abilities. Bowen (2004) suggests that there are benefits in combining TBLL with a more systematic coverage of grammar and vocabulary development to provide a comprehensive approach that can be adapted to meet the needs of all learners.

The advantages of TBLL include the following:

- Learning takes place in an authentic context.
- Students learn from one another.
- Learners gain in confidence.
- It provides an inclusive learning experience for students in mixed-ability classes.
- Skills and concepts learned through TBLL are more easily generalised.

Potential difficulties in TBLL include the following:

- Some teachers (and learners) feel uneasy about having no particular sequence for introducing and studying new concepts and vocabulary from the simple to the complex level.

- Students lacking in confidence may feel inhibited by discussion situations and may be reluctant to participate.

- Students may apply and practise incorrect language patterns or vocabulary without receiving corrective feedback.

Cognitive strategy training

A cognitive strategy (or 'learning strategy') is a mental plan of action that allows a learner to approach a learning task systematically. Using such a strategy facilitates a greater chance of success than would occur simply by trial and error methods. Cognitive strategies are involved in all learning activities that require thinking, planning and decision-making — such as writing an essay, note-taking, researching a project, tackling a mathematical problem or conducting a science experiment. An effective cognitive strategy enables learners to plan what they will do, implement their plan, monitor what they are doing, and modify their thoughts and actions, if necessary, as they proceed.

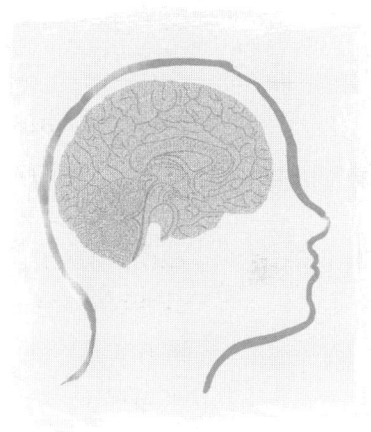

For the past two decades it has become increasingly clear that students differ greatly in the extent to which they are (or are not) 'strategic' learners. For example, students with learning difficulties are often observed to display poor task-approach skills and inefficient learning strategies. These weaknesses can cause or exacerbate a student's learning problems (Brophy, 1998). It has also become clear that cognitive strategies can be taught to students who appear not to have developed them, resulting in improved achievement and increased self-efficacy (e.g. Graham & Bellert, 2005; Van Kraayenoord, 2004). Instruction in use of cognitive strategies has also proved useful in increasing the study skills of adults undertaking training courses (Shenkman, 2002). Together with Direct Instruction (DI), cognitive strategy training has proved to be an effective approach for improving the outcomes for students with learning difficulties, as well as benefiting all other students in the class (Harris, Graham & Mason, 2003; Swanson, 2001).

Ellis (2005) suggests that there are three main types of learning strategy:

- *Cognitive strategies:* These might also be referred to as 'task-approach strategies' and often involve the selection of appropriate procedures to apply when attempting the task at hand. For example, when reading a difficult chapter in a textbook, a student may decide to use the PQRS strategy (preview, question, read, summarise) in order to identify and extract the main points, or may decide to work back from the answer given in the mathematics textbook in order to determine how a problem was solved.

- *Metacognitive strategies:* These involve self-monitoring and 'thinking about one's own thinking' while engaged in a task: for example, 'Am I understanding this? Do I need to go back to the beginning and check?', 'Is this working out OK?' or 'Am I working too quickly?' Metacognition is what enables a learner to modify his or her thinking and actions in the light of the progress being made.

- *Self-regulation strategies:* These are closely related to metacognition, and represent the ways in which a learner may decide to change his or her approach to a problem. For example, a student may decide to seek help from a peer at a particular moment, or check the spelling of a word in the dictionary. Self-regulation in classroom contexts often involves time management so that work is completed, if possible, within the time allotted.

Within the context of various school subjects, direct teaching methods are usually employed to teach cognitive strategies to students who lack them (Swanson 2001; Westwood, 2003a). The teacher will explain the value and purpose of the particular plan of action, and will then 'think aloud' while demonstrating an effective strategy for a particular learning task. The students observe, and develop a similar 'self talk' internal script to help them apply the strategy effectively. Guided practice is then provided, with corrective feedback from the teacher. Students are encouraged to monitor and reflect upon the effectiveness of their own use of the strategy. Generalisation of a strategy to new contexts is always problematic, so deliberate efforts must be made to help students identify other situations in which the same strategy can be used (Van Kraayenoord, 2004).

Advantages of cognitive strategy training include the following:

- Students become more self-regulated, and feel more in control of their own results.

- Students understand the demands of specific tasks more clearly.

- Students usually achieve at a higher level.

- Work completion rate and accuracy improve.

- Students' attitudes toward learning improve.

- Students learn to use tactics such as mnemonic devices, rehearsal, visualisation and self-questioning to aid storage and retrieval of information.

Difficulties involved in teaching and learning cognitive strategies include the following:

- Teachers are not always aware of the need to teach students 'how' to tackle the work they set.

- Unless students are constantly reminded to use the strategies they have been taught, their use will not become automatic.

- It usually takes much longer than teachers anticipate for a learner to adopt a strategy for independent use.

- Strategies are not always taught in a highly relevant context: the teachable moment should be when the students urgently need this knowledge in order to attempt a task.

- For students with learning problems, the strategies taught should not be too complicated and should not have too many steps.

Cognitive apprenticeship

We tend to think of the term 'apprenticeship' as applying to formal training within a profession or a trade, but the teaching and learning principles embodied in apprenticeships can be translated into valuable classroom practices. The notion of apprenticeship implies that a learner acquires new knowledge and skills from an 'expert', partly as a result of direct teaching (instruction, demonstration, practice and feedback) and partly by incidental observation of what the expert does, thinks and says while engaged in a task. In cognitive apprenticeships, learning takes place within the individual's zone of proximal development. It occurs while the learner is working on an authentic task that is slightly more difficult than he or she can manage independently, thus requiring coaching and scaffolding from peers or instructors to succeed (Tretiakov, Kinshuk & Tretiakov, 2003). Gradually the amount of support provided to the learner is reduced (faded) until the individual can complete the task independently and can continue to increase in expertise.

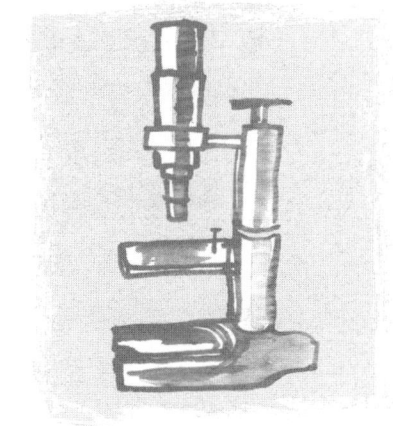

The most important element of cognitive apprenticeship is the passing on of key strategies related to approaching a task effectively and solving problems that may occur (Collins, Brown & Holum, 1991). Conway (1997) has explained that:

> Cognitive Apprenticeship is a method of teaching aimed primarily at teaching the processes that experts use to handle complex tasks. The focus of this learning-through-guided-experience is on cognitive and metacognitive skills, rather than on the physical skills and processes of traditional apprenticeships. Applying apprenticeship methods to largely cognitive skills requires the externalization of processes that are usually carried out internally. Observing the processes by which an expert listener or reader thinks and practices these skills can teach students to learn on their own more skillfully.

In order to teach these task-approach techniques and cognitive strategies to novices, teachers typically use a 'think aloud' approach in which they verbalise their ideas, concerns and plans as they carry out the task. An essential part of the coaching in apprenticeship contexts is therefore teaching the learner 'how to think about' the problem or procedure. For example, Scardamalia, Bereiter and Steinbach (1984) have applied the concept of cognitive apprenticeship to helping students gain more sophisticated skills for writing (composing) through a process of explicit modelling, prompting, coaching, practice, feedback and support. Many of the features of cognitive apprenticeship are also evident in reciprocal teaching (Palincsar & Brown, 1984) — as described in the next chapter — and in situated learning (see pp. 45–46).

There are many different ways in which the apprenticeship approach can be implemented in classrooms, but they all rely to some extent on:

- Using authentic tasks or problems.
- Modelling effective ways of approaching the task or problem (thinking aloud and demonstrating).
- Feedback and coaching for the learner while engaging in the activity.
- Discussion with the learner concerning his or her performance.
- Gradual withdrawal of support as the learner becomes proficient.

Advantages of cognitive apprenticeships include the following:

- Students acquire effective strategies and tactics for attempting many types of task.
- A thoughtful approach to learning is encouraged.
- Collaboration, sharing and support among learners are encouraged.
- Teachers think much more deeply about the cognitive and metacognitive demands of the tasks they set for students.
- When working with lower ability learners, teachers begin to understand their difficulties much more clearly and can provide precise assistance.

Difficulties involved in implementing cognitive apprenticeships include the following:

- Large class size is a major obstacle; implementation is much more viable in small tutorial groups and in one-to-one training contexts.
- Cognitive apprenticeship is not a relevant model for all types of learning. It is useful mainly when a teacher needs to teach students how to tackle fairly complex tasks.

Situated learning

It is said that a weakness of almost all teaching in schools is that it inevitably does not take place in the context where the particular knowledge or skills will be applied in real life. Too much of the information acquired via a traditional curriculum is described as 'inert', because the learner does not recognise its functional value, and is unlikely to recall it later for use at the appropriate time (Alexiades, Gipson & Morey-Nase, 2001). A challenge for educators is to make learning more reality-based and functional; the practical application of the principles of situated learning is one attempt to address this issue. At the moment, situated learning is implemented mainly in upper secondary schools and in colleges or universities. However, some innovative work using similar principles has also been attempted in primary schools.

The common definition of situated (or situational) learning is that it is a form of education that takes place in a real or simulated setting that is functionally identical to that where the learning will need to be applied, such as a workshop, laboratory, kitchen, library or field trip. Alternatively, situated learning principles can be implemented through the process of investigating computer-simulated real-life issues and problems. Computer simulations and 'virtual reality' experiences can be acceptable substitutes for real situations.

The theory of situated learning is an example of the practical application of constructivist learning theory (Tretiakov, Kinshuk & Tretiakov, 2003). Situated learning has links to the experiential learning principles identified many years ago by Dewey (1933), and with notions of scaffolded instruction and social aspects of learning derived from the theories of Vygotsky (1978). An important element in situated learning is social interaction (Brown, Collins & Duguid, 1989). In most real-world learning situations, novices are usually sharing the environment with co-workers, supervisors or experts engaged in the same activity who will help them learn and will respond to their questions and concerns. Many applications of situated learning are therefore closely related to the concept of cognitive apprenticeship (Lave & Wenger, 1991) and to the notion of 'communities of learners' (Brown & Campione, 1994).

According to Vincini (2003), the main advantages of situated learning include the following:

• Learning opportunities are provided in real or simulated contexts in which new knowledge or skills must be acquired for immediate use.

- Experts or mentors are available to provide the learner with support.
- Instructional scaffolding and direct coaching are provided as necessary.
- Tacit knowledge ('know-how') possessed by competent performers is deliberately made explicit for the novice learner.
- Collaboration is encouraged.
- Opportunities are provided for learners to reflect upon and consolidate their learning.

Some of the difficulties associated with situated learning include the following:

- For most teachers, the task of arranging and maintaining real-life learning situations adds considerably to their workload, and also demands ingenuity on their part.
- Assembling computer-simulated resources and links is demanding of time and technical expertise.
- Situated learning is most difficult in schools and systems following prescriptive curricula with an emphasis on examination results.

Anchored instruction

The development of anchored instruction (AI) as a teaching strategy is associated mainly with the work of Bransford and his associates at Vanderbilt University (e.g. Bransford & Stein, 1993). It is described as a technology-based example of situated learning that uses the context of various real-life settings (often simulated) to aid the development of higher order problem-solving skills and to facilitate generalisation and transfer of learning (Booth, 2000). In a typical anchored instruction session, a feasible problem or case study with embedded data is presented in a narrative format, often on video. Students identify with aspects of the problem and become actively involved in generating possible solutions. The scenarios depicted in the videos serve to 'anchor' the learning in a context to which the learner can relate. Learning that occurs as students work through a problem is believed to be more easily transferable to new situations than is inert information gathered from traditional lecture or textbook presentations.

There are no set procedures for developing or using AI, but sessions usually follow this format:

1   Teacher or instructor introduces the anchor problem or case.

2   The group discusses the issues involved, and individuals share their experiences of real-life situations that may relate in some way to working through the problem or case.

3   Students then work on the problem in small groups, drawing on any resources (human or material) required.

4   Groups later share and discuss their possible solutions and offer suggestions to help refine final decisions.

Advantages of AI include the following:

- It is a motivating approach to learning.
- It makes relevant use of current technology.
- Students are more likely to become confident and independent thinkers.
- Learning is likely to generalise more easily to new contexts.

Disadvantages or difficulties include the following:

- Teachers need to have good ICT skills to create resources for programs of this type.
- There are not many ready-made resources for AI available at this time.
- It is difficult to avoid fragmentation of curriculum content.
- The AI approach is more concerned with process than product.
- The teacher needs to be skilled in giving just the right amount of direction and support.

In the following chapter, attention is given to ways in which teachers and students can interact in the classroom during the course of instruction. Regardless of whether a lesson is presented mainly by teacher-directed methods or a student-centred approach, learning is facilitated by the way in which students can communicate and work with the teacher and with each other.

Useful resources

Discovery learning

Jacobs, J. (2004). The limits of 'discovery learning'. Message board, 4 February, joannejacobs.com. Viewed 1 February 2006, (http://www.joannejacobs.com/mtarchives/013751.html).
Some dissenting (or at least cautious) views on the value of discovery learning can be found in this online post and its replies.

Ormrod, J.E. (2000). *Educational psychology: Developing learners* (3rd edn). Upper Saddle River, NJ: Merrill-Prentice-Hall.
This text has a very good section on the principles of DL, together with hints on enhancing its effectiveness.

Resource-based learning

Campbell, L., Flageolle, P., Griffith, S. & Wojcik, C. (2002). Resource-based learning. In M. Orey (ed.) *Emerging perspectives on learning, teaching, and technology* (online book). Athens, GA: University of Georgia College of Education. Viewed 1 February 2006, (http://www.coe.uga.edu/epltt/RBL.htm).

Esch, C. (2005). *Resource-based learning: Guide to good practice.* Southampton, UK: Subject Centre for Languages, Linguistics and Area Studies. Viewed 1 February 2006, (http://www.lang.ltsn.ac.uk/resources/goodpractice.aspx?resourceid=409).

These texts supply more information on RBL.

Project-based learning

Buck Institute for Education (2002). *Project based learning handbook.* Novato, CA: Buck Institute for Education. Viewed 1 February 2006, (http://www.bie.org/pbl/pblhandbook/intro.php).

Houghton Mifflin (n.d.). *Project based learning.* Wilmington, MA: Houghton Mifflin Online Study Center. Viewed 1 February 2006, (http://college.hmco.com/education/resources/res_project/students/background.html).

These texts supply more information on project-based learning.

Problem-based learning

Eggen, P. & Kauchak, D. (2004). *Educational psychology: Windows on classrooms* (6th edn). Upper Saddle River, NJ: Pearson-Merrill.

Excellent coverage of problem-based learning.

ERIC EC (ERIC Clearinghouse on Disabilities and Gifted Education) (2003a). *Problem-based learning.* Arlington, VA: Council for Exceptional Children. Viewed 1 February 2006, (http://ericec.org/faq/gt-prob.html).

Additional information on PBL in school contexts.

Pressley, M. & McCormick, C.B. (1995). *Advanced educational psychology for educators, researchers and policymakers.* New York: Harper Collins.

Pressley and McCormick provide insights into the feasibility of teaching problem-solving strategies and identify some of the potential difficulties.

Planet (2001). Special Edition 2, November. Viewed 1 February 2006, (http://www.gees.ac.uk/pubs/planet/pbl.pdf)

The whole of *Special Edition 2* of the journal *Planet* was devoted to issues of implementing PBL.

Task-based learning

Bygate, M., Skehan, P. & Swain, M. (eds) (2001). *Researching pedagogic tasks, second language learning, teaching and testing.* Harlow: Longman.

Edwards, C. & Willis, J. (eds) (2005). *Teachers exploring tasks in English language teaching*. Basingstoke: Palgrave Macmillan.

These texts supply more information on task-based learning.

McKinnon, M. & Rigby, N. (2004). *Task-based learning*. Basingstoke, UK: Onestopenglish/Macmillan Publishers. Viewed 1 February 2006, (http://www.onestopenglish.com/teacher_support/methodology/archive/teaching-approaches/task_based_learning.htm)

Provides a good summary of TBL.

Strategy training

Pressley, M. (1999). Self-regulated comprehension processing and its development through instruction. In L. Gambrell, L.M. Morrow, S.B. Neuman & M. Pressley (eds) *Best practices in literacy instruction* (pp.90–97). New York: Guilford Press.

Pressley, M. & Woloshyn, V. (1995). *Cognitive strategy instruction that really improves children's academic performance* (2nd edn). Cambridge, MA: Brookline Books.

These texts supply additional information on strategy training.

Sabornie, E.J. & deBettencourt, L.U. (2004). *Teaching students with mild and high-incidence disabilities at the secondary level* (2nd edn). Upper Saddle River, N.J: Merrill-Prentice Hall.

This text contains a useful chapter referring to strategy training for students with special needs.

Swanson, H.L. (2001). Searching for the best model for instructing students with learning disabilities. *Focus on Exceptional Children* 34, 2, 1–15.

Swanson provides details of the effectiveness of strategy training and direct instruction, indicating the complementary nature of the two methods when used together.

Cognitive apprenticeship

Collins, A., Brown, J.S. & Holum, A. (1991). Cognitive apprenticeship: Making thinking visible. *American Educator*, winter issue. Viewed 1 February 2006, (http://www.21learn.org/arch/articles/brown_seely.html).

This text supplies additional information on cognitive apprenticeships.

Situated learning

Lave, J. and Wenger, E. (1991). *Situated learning: Legitimate v peripheral participation*. Cambridge: University of Cambridge Press.

McLellan, H. (1995). *Situated learning perspectives*. Englewood Cliffs, NJ: Educational Technology Publications.

Vincini, P. (2003). The nature of situated learning. *Innovations in Learning*, February. Somerville, MA: Tufts University. Viewed 1 February 2006, (http://at.tccs.tufts.edu/pdf/newsletter_feb_2003.pdf).

These texts supply additional information on situated learning.

Anchored instruction

Foster, C. (2004). Anchored instruction. In B. Hoffman (ed.) *Encyclopedia of Educational Technology*. San Diego: San Diego State University. Viewed 1 February 2006, (http://coe.sdsu.edu/eet/articles/anchoredinstruc/start.htm)

A useful overview of anchored instruction.

Adventures of Jasper Woodbury (1992). Set of 12 interactive videodiscs. Mahwah, NJ: Learning Inc.

A good example of AI can be seen in this interactive videodisc program. Details are available online at (http://peabody.vanderbilt.edu/projects/funded/jasper/).

4 Classroom interactions for learning and teaching

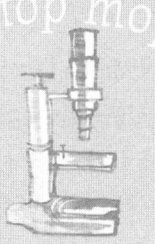

A community of learners is a classroom in which teacher and students actively and cooperatively work to help one another learn; it often incorporates approaches such as class discussions and cooperative learning. Although the teacher provides some guidance and direction for classroom activities, students assume much of the responsibility for facilitating one another's learning (Ormrod, 2000, p.594).

Almost all learning in school involves some degree of social interaction. Students learn not only by listening to a teacher but also by observing, imitating, communicating and interacting with the teacher and with each other. Even in the most highly teacher-directed learning situations, students are watching and listening to the ideas and responses of others, and may ask each other for information or assistance.

In this chapter, we will consider the important roles of attention, questioning, discussion, peer assistance and cooperative group work in increasing the amount of positive interaction and learning in the classroom. In the following chapters that deal with specific curriculum areas, reference will be made again to these variables.

Establishing attention

The first requirement for learning anything new is careful attention to task. In order to acquire new concepts and skills it is necessary for a learner to attend closely to situations in which material is presented or in which information is available to be discovered. Without adequate attention, very little information is effectively processed and transferred from the learner's short-term memory into long-term memory (Ormrod, 2000).

Many learning problems in school are due simply to the fact that the learner was not attending adequately during the lesson and therefore failed to see and hear important information, or failed to carry out the teacher's instructions. Poor attention to task is often cited as a common characteristic of students with learning difficulties (Burgess, 2003; Mastropieri & Scruggs, 1993) and unfortunately the adverse effect of poor attending tends to be cumulative — once a student has missed vital information or failed to master skills over several lessons it becomes even more difficult to pay attention in future lessons because the material is not making sense. Many cases of so-called attention 'deficits' are simply a reflection of the individual's adaptive response to a frustrating situation in which lesson content or task demands are not understood.

It is true that a few students have chronic difficulty maintaining attention to one topic for more than a few moments before they lose focus or are distracted by something else. These students are often described as having *attention deficit disorder* (ADD). Some of these students also display 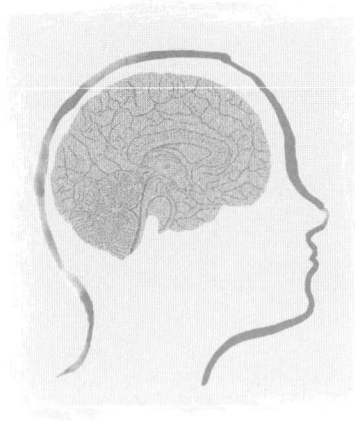 abnormally high levels of impulsive and uncontrolled physical activity, and may be diagnosed by psychologists or doctors as having *attention deficit hyperactivity disorder* (ADHD). Genuine cases of ADD and ADHD are thought to have biological and possibly genetic causes (Barkley, 2003) and they represent no more than three to four per cent of the school population. However, there are probably very many students who are incorrectly given the ADD/ADHD label simply because they exhibit the same inability to attend to lessons and control their behaviour due to frustration, boredom and anxiety related to school work and failure. Helping students with chronic attentional problems may necessitate teaching them self-monitoring techniques so they can determine how well they are attending to the task at hand and modify their own behaviour (West Virginia University, 2005).

Students' poor attention can be due to many intrinsic and extrinsic factors, including anxiety; fatigue; illness; boredom; distractions; language difficulty (or language difference, as in English as a second language situation); poor self-regulation; low motivation; teaching that does not mesh with students' prior knowledge, interests or experience; and subject matter that is overly complex or irrelevant. Young or 'immature' students tend to be distracted easily by the children around them (particularly during group work), by noises and movement from outside or inside, by visitors in the classroom, and so forth. This tendency becomes less for most children as they get older, but a few remain highly distractible. Even the learning environment itself can be a source of distraction: for example, when secondary school students with learning difficulties move from their familiar classroom to the science laboratory, they can become distracted by many interesting pieces of equipment and the increased opportunity to touch things. The same situation applies when distractible students are on field trips outside school.

One of the common recommendations for teaching students with attention problems is to reduce to a minimum any potentially distracting stimuli in the learning environment, and to avoid too many transitions and disruptions during the lesson. But, in reality, this advice is not easy to implement in a typical classroom. More realistically, Burgess (2003, p.9) wisely states, 'Sound teaching practices for students with difficulties attending to classroom activities require direct instruction or explicit teaching when introducing topics or specific skills'. Such an approach, involving pedagogy rather than modifying the environment, is much more manageable for teachers.

Gaining and maintaining students' attention usually requires the teacher to apply several of the following strategies:

- Spending enough time to settle students down and get them focused on the topic of the day at the beginning of every lesson.
- Showing personal enthusiasm for the topic (because teachers' enthusiasm plays a major role in gaining students' interest).
- Waiting to begin the lesson until the attention of every student in the room is established.
- Maintaining frequent eye contact, particularly with students who tend to daydream or who are easily distracted.
- Waiting to pass out materials until students understand fully what they are required to do.
- Making effective use of visual aids (real objects, models, charts, apparatus, video, DVD, CD-ROMs, Internet and relevant computer software).
- Talking less and listening more (because a purely chalk-and-talk lesson can exceed the concentration span of *all* the students).
- Encouraging students to talk and contribute more.
- Making practical demonstrations clear, and checking for understanding.
- Communicating clearly and effectively.
- Checking for understanding at every step in the lesson.
- Having sufficient practical activities to engage all students.
- Assigning students only one task at a time to avoid confusion.
- Asking many relevant but easy questions.
- Stepping in and supporting any student who seems to be having difficulties.
- Avoiding dead spots in the lesson and avoiding loss of time in transition from one activity to another.
- Avoiding seating two students together who are likely to distract one another.
- Regularly calling the class to order, and reviewing or consolidating what has been done.

When teachers are skilled in establishing and maintaining students' attention, they are well on the way to establishing an effective learning environment. As Evans (1995, p.247) has remarked, 'Without attention, there can be no learning'.

Questioning

Regardless of the overall teaching method used in lessons, effective questioning of students is a vital instructional strategy and one that is used widely (DfES, 2004; Kauchak & Eggen, 2003; Ritchie, 2001). Even when a teacher is explaining  something to the class by direct methods, it is not normally a one-way communication process. The teacher asks questions of the students as the lesson progresses to involve them more fully in the work and to ensure that what is being presented is making good sense. High rates of questioning by the teacher are naturally associated with high rates of responding by the students — as, for example, in interactive whole-class teaching and direct instruction (see Chapter 2). In turn, these higher rates of response are associated with higher achievement (Cole & Chan, 1987; Stigler & Hiebert, 1997). Teachers in classes showing high achievement levels are usually found to ask many questions during a lesson, with a low proportion of those questions yielding incorrect responses or no response from the students. Effective teachers also actively encourage students to ask questions during and after an explanation.

Teachers' questioning generally serves the following purposes:

- To check what students already know or believe before a new topic is presented.
- To engage students' curiosity and to arouse interest.
- To focus attention on salient aspects of a learning situation.
- To give positive direction to students' thinking.
- To promote reflection on the significance and value of observations.
- To challenge and extend logical reasoning.
- To lead students through a planned learning sequence.
- To review a topic and facilitate recall.
- To aid transfer and generalisation by helping students forge appropriate connections with other subjects and with real-life applications.
- To check for understanding and to identify misconceptions.

It has been demonstrated that children with poor learning skills seem to benefit most from instruction that includes a high percentage of simple direct questions focusing on the core content of the lesson (Brophy & Good, 1986). It is as if answering these questions helps firm up the student's understanding of the topic. Most core questions are referred to as 'lower order' questions, and it has been suggested that about 80 per cent of classroom questions should be of this type. If students are struggling to assimilate basic facts, it is usually necessary to ask many lower order questions. On the other hand, if development of critical thinking is the target, higher order questions are necessary (Good & Brophy, 2003). Bell (1999), writing mainly about teaching science to intellectually disabled students, suggests that a teacher needs to be skilled in asking questions that do not result in an immediate response of 'I don't know'. To this end, it can be beneficial to break challenging questions down into a series of much easier questions that build gradually toward the higher cognitive level.

An important aspect of questioning that has been studied is what is called 'wait time'. Rowe (1978; 1986) analysed audio tapes of lessons and discovered that teachers often asked between three and five questions a minute, but allowed only a second or so for a student to respond before asking someone else or providing the answer themselves. It was found that when teachers deliberately extended wait time to three seconds or more after they asked a question *and* after a student responded, the following things occurred:

- The number of responses increased.
- Fewer instances of failure to answer occurred.
- Students' confidence in answering increased.
- The length of a student's response increased.
- The number of questions asked by the students themselves increased.
- Contributions from lower ability students increased.

When working with students with learning difficulties it is important to recognise that some of them are afraid to speak up or give opinions because they are fearful of being wrong (Bell, 2002). A very important way in which teachers can address individual differences among students is to differentiate the level of complexity in their questioning — that is, to direct simple questions at first to the less confident students, and more challenging questions to other students. If students with learning difficulties are asked fairly basic and direct questions in the beginning they are more likely to offer a response, feel successful and remain engaged in the activity. Unfortunately, students with learning difficulties are the least likely to ask spontaneous questions or to seek clarification from the teacher or peers (Good & Brophy, 2003). Skilled teachers will notice this fact and try to draw such students into the lesson by directing questions to them at an appropriate level of difficulty. Some common errors that teachers make when questioning include:

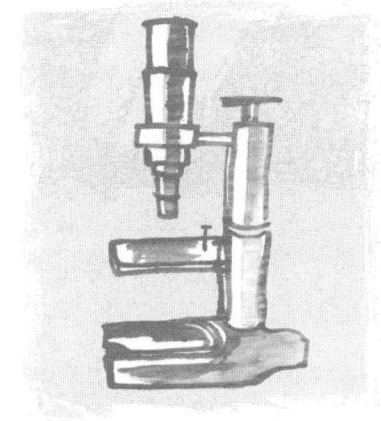

- Asking too many difficult or poorly worded questions, thus causing confusion.
- Continuing to ask questions on a topic even though the students have indicated clearly a lack of knowledge of the topic.

- Taking answers only from students who volunteer.

- Not allowing adequate wait time for students to formulate answers.

- Always insisting that an individual student gives an answer, rather than allowing some answering by the class in unison.

Discussions

Discussion between a teacher and students (and among students themselves) is an effective teaching strategy in most curriculum areas. According to Kauchak and Eggen (2003), effective discussions in the classroom can stimulate students' thinking, strengthen their understanding of the topic and challenge their existing attitudes and beliefs. To be effective, discussions must have a very clear purpose and must be well-managed.

For discussions to be useful, students must have sufficient background experience and knowledge to enable them to contribute. For example, in secondary schools, students may have been required to read a particular chapter of the textbook or of a novel for homework, and to come prepared to give their views on certain issues. The teacher may even set the framework for discussion by putting some leading questions on the whiteboard. Without this background knowledge all members of the group would simply 'share their combined ignorance' and rapidly lose interest. In such situations, behaviour can deteriorate and problems can arise.

Effective discussion, whether conducted in whole-class or small group format, also requires that the teacher monitor closely what is occurring throughout the session. At fairly frequent intervals it is important to pause and call the class together to review ideas emerging from the discussion. The teacher can then clarify any points of confusion or doubt, and refocus the students on the remaining issues. Time management is important so that there is an opportunity at the end of the session to summarise and consolidate key ideas from the discussion and share these as the closure for the lesson. Barry (1995, p.389) suggests that the role of the teacher can best be described as 'chairperson, guide, initiator, summariser and referee'.

Small group discussions ensure that opportunities are increased for students to participate and contribute their ideas. Small groups, however, are potentially more difficult to manage from the teacher's point of view, and some groups may well get off task before the teacher is able to redirect them or provide necessary input and support. This problem is exacerbated when the teacher has not been entirely clear in stating his or her expectations when setting appropriate ground rules and goals for the discussion.

Problems that can occur with discussions include:

- Students are not completely clear about what they are required to do.
- Time is managed badly, so that no closure and summarising are possible.
- A small number of assertive students dominate the discussion.
- Students lack the requisite interpersonal skills to engage in discussion without friction or argument (for example, an unwillingness to listen to the views of others or to compromise).
- Errors of fact and misconceptions are not detected and corrected by the teacher.
- The teacher gives too little support, encouragement and direction during the session.
- The topic is of no real interest to the students.

It is generally accepted that classroom discussions are valuable if conducted efficiently, and probably do stimulate students' thinking as well as facilitating deeper processing of lesson content. However, Pressley and McCormick (1995) point out that effective discussions are not easy to organise, and there is no guarantee that they will go well and curriculum content will be learned. Discussions should not simply become a routine that is introduced for a time into every lesson. There must be a clear purpose in requiring students to engage in this activity in order to achieve specific objectives.

Some of the most valuable discussions that occur are when two students are working together on a common task, as for example in peer tutoring or cooperative learning. Working with a partner maximises the opportunity to engage in transactional talk.

Same-age peer tutoring and cross-age tutoring

It is widely believed that one-to-one teaching is far more effective than group teaching (Brown, Morris & Fields, 2005; Gaustad, 1993). This fact was recognised by Bloom (1984), the originator of mastery learning, who challenged educators to try to devise group methods of instruction that could come anywhere close to producing the same achievement results that occur from individual tuition. Even now, with the use of computer-assisted instruction, this ideal has not been fully achieved.

The benefits of individual tuition mainly stem from the way in which the learner's attention is held and instruction can be geared closely to the individual's present capabilities. The tutor is able to detect misunderstandings as they arise and reteach, demonstrate or explain again the material in a different way

(Kauchak & Eggen, 2003). Individual tutoring also increases the opportunity for a learner to ask many questions of the tutor, to get immediate feedback and to make many more responses during the session than can occur under group or whole-class teaching methods (Arreaga-Mayer, 1998; Graesser & Person, 1994).

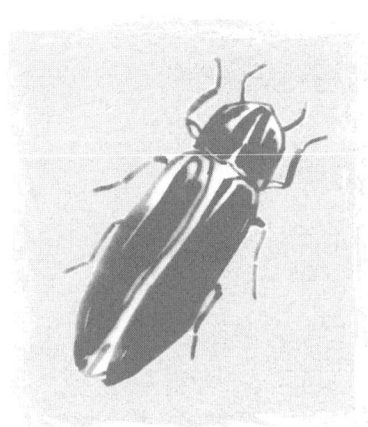

Unfortunately, most teachers rarely have an opportunity to provide intensive one-to-one tuition in the classroom. They may give fleeting individual attention to a student who is having difficulties, and will perhaps respond to questions students raise; but, in general, the need to manage the larger group and adhere to the time plan for the lesson prevents teachers from combining whole-class processes with much individual attention. The compromise is obviously to empower students in the class to help one another via peer tutoring. Students in the class are a teacher's most readily available human resource for supporting learning. Having students assist each other to master lesson content and to correct misunderstandings appears to be a very effective way for teaching and practising fundamental knowledge and skills (Ormrod, 2000).

The term 'peer tutoring' is usually applied in situations when the student who is doing the tutoring is the same age as the student being tutored. Sometimes schools arrange for older students to help younger students, in which case the term 'cross-age tutoring' is applied (Gaustad, 1993). When all students in a class are involved in assisting one another, the arrangement is referred to as 'classwide peer tutoring' or CWPT (Greenwood, 1991). It is claimed that CWPT can triple the amount of practice time that students engage in, compared with more traditional teacher-led lessons (Greenwood & Delquadri, 1995). Fuchs, Fuchs and Burish (2000) have reported favourably on a similar arrangement, termed *Peer-assisted Learning Strategies* (PALS), using learning partners and targeting reading and mathematics in Years 2 to 6.

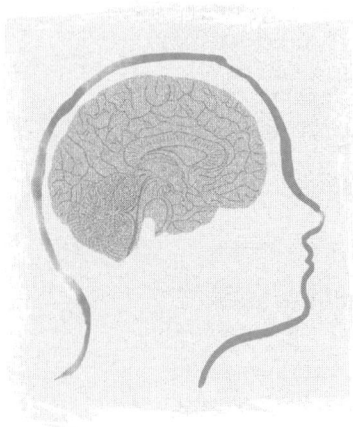

Research supports the view that peer tutoring and cross-age tutoring are relatively effective in improving both tutees' and tutors' academic and social development (e.g. Greenwood, Delquadri & Hall, 1989; Maheady, Harper & Mallette, 2001). Peer and cross-age tutoring have also proved valuable as strategies for helping students with special needs in inclusive classrooms and increasing their time on task (Kavale & Forness, 2000; Serna & Patton, 1997a). Studies involving students with various categories of disability have found self-esteem, attitudinal and social benefits as well as academic gains in reading and basic mathematics. Kalkowski (2001, p.2) has reported that:

Research on low-achieving and other high-needs students as tutors has increased in the last decade. Both wide-ranging reviews and individual studies show impressive gains for low-achieving, limited-English-speaking, learning disabled, behaviorally disordered and other at-risk student populations in both the academic and affective realms and at all age/grade levels.

Acting as 'instructor' and 'guide' does not come naturally to every student. For this reason, advocates of peer and cross-age tutoring agree that tutors need to be trained or briefed to some extent so that they can carry out the role effectively. In particular they may need help in breaking a task down into easier steps, giving corrective feedback, listening more and talking less, praising and encouraging the tutee. Much good advice on such matters can be found in the books and papers by Topping (e.g. Topping, 1995; Topping et al., 2000). Research studies have demonstrated clearly that with proper training, students can successfully tutor other students with excellent results (Gaustad, 1993; Yamanashi, 2005). One reason for this may be that students working together communicate more effectively ('speak a similar language') than do teachers and students, and the learning situation is more relaxed (Damon & Phelps, 1989).

The benefits of peer and cross-age tutoring include:

- Improvement in basic academic skills.
- Enhanced social behaviours.
- Increases in self-esteem and self-efficacy.
- More positive attitudes toward school.
- More time spent on task.
- Adapting tutoring to learner's rate and level of understanding.
- Immediate feedback.
- Abundant opportunities to practise.
- Often benefits the tutors as much as their tutees.

Potential difficulties in using peer and cross-age tutoring include the following:

- Some student tutors may not completely understand the material to be taught.
- Some students tend to resist being tutored by peers because they feel it identifies them as stupid and incapable.
- Gains in learning may not be maintained if there is no further support provided.
- Parents of student tutors may feel worried that their child's school time is taken up teaching others rather than learning new material.

- There may be too few tutoring sessions per week to have any major impact.

- It is difficult to arrange the timetable to allow for peer tutoring, and even more difficult to schedule cross-age sessions.

- Some schools are unwilling to attempt the logistical challenge of arranging and supporting such tutoring.

Cooperative learning

Although there are many different models of cooperative learning, they are all characterised by groups of students working together to maximise their own and each other's learning. It is reasonably common to find cooperative group work included as one approach to teaching and learning in both primary and secondary schools. Ormrod (2000, p.577) observes that:

> When students help one another learn, they create scaffolding for one another's efforts, and they jointly construct more sophisticated ideas and strategies than any single group member might be able to construct alone.

Cooperative learning has been comprehensively evaluated over the past 30 years and much of this work has been conducted by Slavin (1995) and by the  brothers Roger and David Johnson (e.g. Johnson & Johnson, 1999; Johnson, Johnson & Holubec 2002). Johnson, Johnson and Stanne (2000) collated a good overview of the research and provided data from a detailed meta-analysis of 158 studies representing different models of cooperative learning. They concluded that all cooperative models had a significant positive impact on student achievement and on affective outcomes. It is claimed that cooperative learning approaches, when efficiently implemented, can produce higher student achievement than can competitive approaches or students working independently. When it is poorly implemented, however, the results from cooperative group work can be inferior to the outcomes from traditional whole-class teaching. It is also true that certain learning objectives are more effectively achieved by using teacher-directed methods (e.g. Shupe, 2003). Good and Brophy (1997, p.284) consider that:

> It is important to view cooperative learning approach not as a wholesale replacement of traditional whole-class instruction but as an adaptation of this approach in which active whole-class instruction by the teacher is retained but many follow-up practice and application activities are accomplished through small-group cooperation rather than through individual seatwork.

For cooperative learning to be effective, students can't simply be put into groups and assigned a task; all members of the group must understand the purpose of the activity, must have a definite role to play in completing the work and must apply social skills of working collaboratively and supportively with others (Serna & Patton, 1997a; Yamanashi, 2005). The success of the group assignment depends on the successful work of every individual member of the group. Johnson, Johnson and Holubec (2002) indicate that effective cooperative learning should involve a feeling of *positive interdependence* ('we all need to work together and help one another'), *individual accountability* ('each of us has to contribute and learn'), *interpersonal skills* (trusting, communicating and resolving conflict), *face-to-face interaction* (discussing, questioning, suggesting, exchanging ideas, making plans), and *reflection* (thinking back on how well the group has functioned and how to improve). It is also essential that all members of a group have an opportunity to contribute and to be valued.

Cooperative group work is regarded as one essential strategy for enhancing the inclusive nature of the classroom program, and is therefore thought to benefit students with special educational needs or learning difficulties (e.g. Westwood, 2003a; Yamanashi, 2005). However, it is true that the communication difficulties or underdeveloped social skills of some students with disabilities can present a major challenge for both teacher and peers during group work activities. Careful thought will be needed to plan the most effective ways of integrating students of very low ability into working groups, perhaps by starting with a partner ('buddy system') and moving later to inclusion in a slightly larger group.

The advantages of cooperative learning include:

- Application of socio-constructivist principles of learning.
- Active involvement of students.
- Promotion of higher academic achievement.
- Stimulation of independent thinking and cognitive development.
- Boosting students' skills in oral communication.
- Motivation of students to learn the material.
- Developing social and group skills beneficial for success outside the classroom.
- Facilitating opportunities for students of different abilities and different background experiences to work together.
- Enhancing students' satisfaction with their learning experiences.
- Promotion of self-esteem.

Potential difficulties involved in cooperative learning include the following:

- Some students waste time because they are confused or they can't locate necessary resources.
- Some members of the group do not get along well together, so arguments arise.
- Some students do not like group work, preferring to work on their own.
- Behaviour problems can arise.
- There may be an excessive noise level in the room.
- Managing several groups effectively and responding to their needs is difficult for the teacher.
- Some students (particularly those who are not confident or assertive) will opt out and not contribute to the work.
- When the result produced by a group is not good, there is a tendency for other members to blame the lower ability students.
- The teacher may not be effective in consolidating the final outcomes from the group work.

Reciprocal teaching

Reciprocal teaching (RT) is a group learning method devised by Palincsar and Brown (1984) to help students increase their ability to comprehend and interact with text. While RT is mainly associated with the teaching of reading, it is not restricted to 'language arts' and can be used as a teaching strategy in any subject area where students are required to access information via the print medium.

RT combines aspects of explicit instruction with peer teaching, cooperative learning and cognitive apprenticeship. The comprehension strategies are taught first by the teacher, and include questioning, clarifying, predicting and

summarising. After the teacher has described and demonstrated ways of using these strategies before, during and after reading a text passage, the students practise the same strategies, taking it in turn to lead the group in reading, interpreting and reflecting upon information contained in the text. Initially, reciprocal teaching can be introduced to a whole class, but as the students become more experienced and confident in taking over part of the role of a teacher they work more independently in smaller groups. In groups of four, the students can take turns to carry out the roles of questioner, clarifier, summariser and predictor. The teacher still has an important role in monitoring these dialogues conducted within the groups, providing input when appropriate and praising students for their ideas and contributions. The long-term aim is to have students master the reading comprehension strategies for their own independent use across a variety of contexts.

The research findings on the efficacy of reciprocal teaching are positive (Rosenshine & Meister, 1994) and in some studies the method, or variations of it, have resulted in measurable improvement in reading comprehension. However, RT is not a particularly easy method to use, and the teacher needs firm classroom control. Students must have a supportive attitude toward one another and good rapport with the teacher if the learner-centred aspects of the approach are to work productively. RT works best when the teacher knows the class extremely well and uses the approach frequently enough for students to become competent and confident in its implementation and management.

Useful resources

Questioning

Good, T.L. & Brophy, J.E. (2003). *Looking in classrooms* (9th edn). Boston: Allyn & Bacon.
Good and Brophy include an excellent section on questioning techniques.

Wragg, E.C. & Brown, G. (2001a). *Questioning in the primary school.* London: Routledge-Falmer.
Wragg, E.C. & Brown, G. (2001b). *Questioning in the secondary school.* London: Routledge-Falmer.
For more information on questioning techniques.

Peer tutoring

Intervention Central (2004). *Kids as reading helpers: A peer tutor training manual.* New York: Intervention Central. Viewed 1 February 2006, (http://www.interventioncentral.org/htmdocs/interventions/rdngfluency/prtutor.shtml).
Special Connections (2005). *An introduction to classwide peer tutoring.* Lawrence, KS: Special Connections/University of Kansas. Viewed 1 February 2006, (http://www.specialconnections.ku.edu/cgi-bin/cgiwrap/specconn/main.php?cat=instruction&section=main&subsection=cwpt/main)
Contain useful advice for setting up a peer training program.

Cooperative learning

Classroom Compass (1998), *1, 2.* Autumn issue. Viewed 1 February 2006, (http://www.sedl.org/scimath/compass/v01n02/welcome.html).
This edition of *Classroom Compass* focuses on cooperative learning.

New Horizons for Learning (2005), *Cooperative learning*. Seattle: New Horizons for Learning. Viewed 1 February 2006, (http://www.newhorizons.org/strategies/cooperative/front_cooperative.htm).

Information and resources on cooperative learning.

Reciprocal teaching

Jones, R.C. (2001). *Strategies for reading comprehension: Reciprocal teaching*. Winston-Salem, NC: ReadingQuest/Wake Forest University. Viewed 1 February 2006, (http://curry.edschool.virginia.edu/go/readquest/strat/rt.html).

Practical suggestions for implementing reciprocal teaching.

5 Teaching basic academic skills: literacy

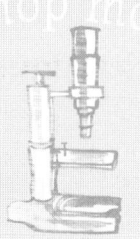

School failure is characterised by low literacy and numeracy levels, alienation, and with dropping out of school. These characteristics are also associated with many secondary students with learning difficulties (Watson & Boman, 2005, p.44).

The ability to read, write and work with numbers ranks perhaps as the most essential learning outcome from a student's years in school. These 'tool subjects', as they were once called, enable students to advance their learning in all areas of the curriculum, and become more independent learners. Literacy and numeracy also allow an individual to remain aware of current world issues and participate actively in the community life of a democratic society (Morcom, 2005).

Unfortunately, some students do not learn to read and write with adequate proficiency, despite being exposed to conventional teaching methods in the primary school years. In Britain, for example, it is reported that approximately 20 per cent of students at age 11 are failing to achieve adequate success in reading and writing at the level expected of students of that age (House of Commons Education & Skills Committee, 2005). Similar problems have been identified in other countries, even where the general standard of literacy may be quite high; for example, Australia (DEST, 2005) and US (National Reading Panel, 2000).

Problems in acquiring literacy and numeracy impact severely on a student's progress in almost every aspect of life, both in and outside school. Failure in basic academic skills usually exerts a detrimental influence on a student's personal and emotional development. While a few students appear resilient to experiences of failure, most students have a low threshold for surviving emotionally such events, and it is common to find them exhibiting signs of low self-esteem, diminished confidence, reduced motivation and increased anxiety (Lancaster, 2005; Martin & Marsh, 2003). Behaviour problems, absenteeism and a 'poor attitude' to school are frequent accompaniments to learning failure.

This chapter will explore briefly some of the teaching approaches thought to be effective in developing students' basic literacy skills and in remedying difficulties in learning. The following chapter will address the teaching of numeracy skills.

Teaching students to read

There are three main approaches to teaching reading — the skills-based approach, the meaning-based approach, and the more recently named 'balanced' approach that combines elements of both skill development and comprehension (DEST, 2005; Ellis, 2005; Tompkins, 2006). The relative merits of skills-based and meaning-based approaches have been debated hotly over many years.

In skills-based reading instruction, students are directly and explicitly taught how to process written language by using phonic knowledge and are given much practice in acquiring this skill. Often the vocabulary-controlled reading material they are given in the early stages is designed to provide abundant opportunities for them to exercise decoding and word recognition skills successfully, and develop a high degree of automaticity (Carnine, 2004). Students are also taught comprehension skills and provided with many opportunities to read for meaning and pleasure.

In the meaning-based approach, teachers give much less attention to direct teaching and drilling of component reading skills such as decoding and spelling, and instead adopt a more holistic method wherein they expose students to a range of interesting 'real' books and motivate children to want to read. The teacher provides some degree of guidance and support to students as they experiment with print and discover for themselves how to read, largely by incidental learning and imitation. Advocates for whole language believe that the reading process is driven not by letter and word recognition (lower order skills) but by understanding the meaning of what is being read (higher order cognition) and by using contextual cues. Castles (2005) points out, however, that although this may be the way that experienced and skilled readers operate with print, it is not how the vast majority of children first learn to read. Most beginners do need to be taught how to decode unfamiliar words. Whole language principles are also embodied in methods known as *language-experience approach*, *shared book experience* and *guided reading* (for details, see Westwood, 2001); but in each of these methods some degree of direct teaching of skills is usually incorporated.

The 'balanced' approach to reading that is currently favoured (DEST, 2005; Hoffman, Baumann & Afflerbach, 2000; House of Commons Education & Skills Committee, 2005) combines the motivating features of whole language (reading for authentic purposes; reading 'real' books with exciting stories) with essential teaching of decoding skills and strategies for comprehending text. Ellis (2005, p.45), having reviewed the research, concludes:

> Consequently, rather than advocating a single-strategy approach, an increasing number of educators are promoting the benefits of balancing student-directed approaches with teacher-directed approaches in the classroom.

Concern has been expressed in both Britain and Australia over the issue of whether pre-service teacher education courses are providing adequate training to enable teachers to adopt this balanced approach when teaching reading (DEST, 2005; House of Commons Education & Skills Committee, 2005). Often, it seems, trainee teachers are exposed only to the meaning-based (whole-language) methodology for literacy development, and they are therefore uncertain how best to embed the teaching of specific skills such as phonic decoding and spelling within early reading programs. It is to be hoped that greater attention will be given in the coming years to ensuring that all beginning teachers are equipped with a broader repertoire of instructional strategies, particularly those relating to the teaching of phonic skills.

Beginning to read

At its most basic level, reading involves two key processes: word identification and comprehension. For students with learning difficulties both processes can be problematic. A significant number of students experience chronic difficulties acquiring swift and automatic word identification, and they therefore understand very little of what they attempt to read. Teaching methods for early reading must equip students with strategies for recognising or decoding unfamiliar words, as well as focusing on the meaning of the sentences.

Most children begin to read by recognising a few words by their visual characteristics; for example, *Goldilocks* is a longer word than *bears* and has a capital letter at the beginning. The word *EXIT* is almost always seen in capital letters in the environment, and the uncommon letter *X* makes it an easy word to remember and recognise. These are the types of visual cues used by students operating at a 'pre-phonetic' or *logographic* stage of literacy development. In the 1960s and 1970s, the popular method of teaching reading (the 'whole-word' or 'look-and-say' method) relied on children building up an extensive vocabulary of words by sight rather than by applying phonic knowledge to decode them. But there is a limit to the number of words a person can store simply as visual images in long-term memory; 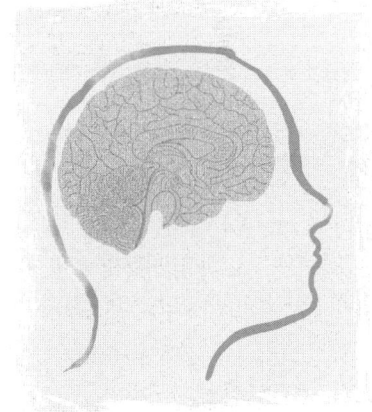 soon a child must acquire a system for tackling new words when reading and writing (Castles, 2005). That system is *phonics*; and learning phonic skills depends upon first mastering a more basic set of underlying *phonological skills*.

Word identification in English language involves obvious skills of visual discrimination and visual memory, but also involves phonological awareness and phonic skills. Phonological skills important for reading include, for example:

- *Segmentation:* The ability to identify separate sounds (phonemes) in spoken words by mentally 'stretching the word' (e.g. *grid*: gr – id; g – r – i – d). This ability to separate out the sounds in words underpins a student's understanding of how the phonic decoding system operates in an alphabetic language. Phonemes are the basic sound units of speech. Each phoneme is represented in print by a letter or group of letters (graphemes). In English, there are approximately 44 such phonemes, and learning phonics means mastering the letter combinations that represent these sounds.

- *Blending phonemes:* This is the reverse of segmentation. Blending involves putting phonemes together to produce a word (e.g. st – o – p: *stop*; bl – a – ck: *black*). Blending sounds is the principle behind what is termed the *synthetic phonics approach* to teaching reading (see p. 69).

- *Detecting words that rhyme* (e.g. *shop, stop, mop, top, hop, pop*): Understanding rhyme helps a student appreciate that certain words share specific clusters of letters, and that these letter groups equate with a sound unit larger than a single phoneme. Learning to recognise letter clusters is the next step in moving from the simple alphabetic (single letter) stage of decoding to the orthographic stage where groups of letters together are recognised as pronounceable parts of words. Understanding rhyme also facilitates the use of analogy for word identification purposes; for example, if I can read and spell the word 'matter' I can probably easily read and spell 'platter'.

- *Identifying initial and final sounds* in spoken words (e.g. knowing that *teacher* and *table* begin with the /t/ sound, and that *back* and *stick* both end with the same sound): Initial sounds and the letters associated with them are extremely important cues when attempting to read unfamiliar words within the context of a meaningful sentence (Castles, 2005).

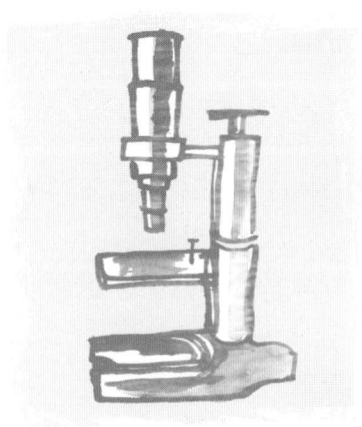

Although some children will have acquired good phonological awareness through informal learning in the preschool years, there will be others who start school with no functional awareness of how separate speech sounds can be identified and used for reading and spelling. Young children who lack adequate phonological awareness are seriously at risk in learning to read and write. Effective teaching of beginning reading usually requires time to be devoted to phonological training activities, as well as the teaching of letter-to-sound correspondences (Tunmer et al., 1998). Such training can be accomplished in part by playing listening games with young children, and partly by direct teaching. Listening games would include activities to introduce and practise word segmentation, sound blending, rhyme and sound identification.

Phonic skills

The teaching of phonic knowledge is best achieved by direct and explicit instruction. Research studies support the view that students will only reach the required level of automaticity in letter–sound correspondences for reading and spelling if they are taught the information thoroughly, and if they practise applying it for the purpose of reading meaningful material (Adams, 1990; DEST, 2005; Ellis, 2005).

Words that at first need to be 'sounded out' are soon recognised easily by sight without decoding, and are added to the reader's 'sight vocabulary'. As readers become more proficient they move to an *orthographic stage* of word recognition. At this stage they are skilled in recognising and using groups of letters that represent pronounceable parts of words, such as *-eed*, *-ence*, *-tion*, *-ite*, *dis-*, *pro-*, etc. Skilled reading is based upon extremely rapid visual processing of such orthographic units. Once competence is achieved, phonic skills become relatively less important because a reader knows many words by sight, and uses clues provided by the sentence context and grammar to predict unknown words.

Research suggests that the teaching of phonics skills should be achieved early in the child's school life, soon after the age of five, to provide a firm foundation on which to build higher order literacy skills (House of Commons Education & Skills Committee, 2005; National Reading Panel, 2000; Rose, 2005). Currently, the favoured method for teaching decoding is to use a *synthetic phonics approach*. This approach is now recommended in the National Literacy Strategy in UK. Rose (2005, p.11) states:

> Because our writing system is alphabetic, beginner readers will not become skilled and fluent, comprehending readers and writers if they cannot understand and operate the system. The case for systematic phonic work is therefore overwhelming and much strengthened by the principles underpinning a synthetic phonic approach.

A longitudinal study involving some 300 children in Scotland (Johnston & Watson, 2005) has given great support for synthetic phonics instruction in Primary 1. The researchers discovered that gains the children made in word reading in Primary 1 increased six-fold by the end of Primary 7, advancing them to three years and six months ahead of their chronological age even though training had ceased after Primary 1. The equivalent gain in spelling was from seven months to one year and nine months ahead of chronological age. Johnson and Watson (2005) point out that this continuing advantage after training stops is very unusual, because the significant effects of specific methods usually diminish rather than increase over time. They suggest that early instruction in synthetic phonics helps children acquire a technique that they can continue to develop for themselves. Synthetic phonics could almost be regarded as a metalinguistic approach in that it provides the learner with insights into how a speech-to-writing language system operates.

In contrast to synthetic phonics approach, a method called *analytic phonics* works in the opposite direction. Analytic phonics teaches children basic letter–sound relationships by analysing words they already know by sight. For example, the sound of the letter *p* is taught from words like *pin, pop, pat* and *pet*. Or, the consonant blend *bl* might be taught from the words *blue, black, blood* and *blink*, and so forth. Analytic phonics is favoured as a method by some teachers because it moves in the direction from meaningful whole to part, rather than part to whole. However, research has not shown it to be more effective than synthetic phonics.

Building sight vocabulary

Of course, there are some very irregular words in the English language that defy phonic decoding. Such words do have to be remembered and recognised as visual images within an individual's sight vocabulary. Even some very common words are difficult to decode and need instead to be mastered by the whole-word method (e.g. *ask, said, are*). A great deal of practice is needed if irregular words are to be stored and recalled effectively. Students must see, read, say and write the words many times to ensure automaticity in recognition. Most of this practice comes from the everyday reading of meaningful text and writing of meaningful communications, but there is also a place for work with flashcards, word games and computer software to help establish word recognition, especially for students

in remedial reading programs (Neill, 2005). For some students with a specific learning disability (dyslexia), or for students with mild intellectual disability, multisensory methods (VAKT) may be helpful in establishing letter and word recognition (see Chapter 2).

If a student has limited sight vocabulary and is overly dependent on phonic decoding, he or she will remain a slow and hesitant reader, experiencing great difficulty understanding what is read. Alongside any structured teaching of phonic skills children must be exposed to methods that ensure high-frequency words are recognised instantly. These words comprise more than 60 per cent of the vocabulary encountered on any typical page, and instant recognition of the words greatly enhances reading fluency. In turn, fluency has a strong positive influence on reading comprehension (Pikulski & Chard, 2005).

Proficient readers are able to make use of several cueing systems in addition to phonics and whole-word recognition to help identify words on a page of text. For example, if the reader understands what he or she is reading, the meaning provides major support for predicting words that are coming next in a sentence: 'I stepped into the lift and pressed the b _ _ _ _ n for the 6th fl _ _ r'. Contextual cues and syntactic cues (the grammar and word order of the sentence) together with initial letter of the next word are powerful aids to word identification and fluency. Explicit methods used to teach reading must ensure that students understand these within-text cueing systems and can apply them appropriately. In other words, students must be taught directly how to make effective use of context.

Comprehension

The primary purpose of reading is to extract meaning from text — indeed, reading could be defined in precisely those terms. Understanding text involves, among other things, the ability to identify words, attach correct meaning to those words, relate the ideas to prior knowledge and keep the train of thought active long enough to process the information in short-term working memory. Comprehension also involves the recall of facts, identification of main ideas, inference, prediction, evaluation and drawing of conclusions. As Rubin (2000, p.171) correctly states, reading comprehension is 'a complex intellectual process involving a number of abilities'.

Sometimes comprehension difficulties arise simply from the student's limited vocabulary knowledge. If there is a serious mismatch between a student's own expressive and receptive vocabulary level and words used in the text, the student will obviously have difficulties understanding. According to Lubliner and Smetana (2005), students with good vocabularies find reading easier, read more widely and do better in school, while children who enter school with a limited vocabulary find reading difficult, resist reading, learn fewer words and consequently fall further behind. In this situation there is a need to devote more time to word study and vocabulary-building when comprehension activities are used in the classroom. There is also a need to preteach difficult vocabulary to such children before a textbook is read.

Proficient readers with efficient word identification skills, a rich vocabulary and good background knowledge can accomplish comprehension of text fairly easily. They have already mastered lower order processes such as decoding and using contextual cues, and they can devote all their cognitive effort to derive meaning from what they are reading. Skilled readers tend to have a number of effective strategies available for use that help them identify main ideas, anticipate what may be coming next, generate questions and reflect upon or appraise the information they are reading. In contrast, Graham and Bellert (2005, p.71) state that:

> Students with learning disabilities often experience poor comprehension due to their failure to read strategically and to spontaneously monitor their understanding while reading.

As students get older they are faced, in secondary schools particularly, with textbooks that are often very demanding in terms of reading ability; and students who lack comprehension strategies are at risk of failure and frustration (Hummel, 2000). Matching the readability level of books to students' current reading ability can do much to increase their comprehension.

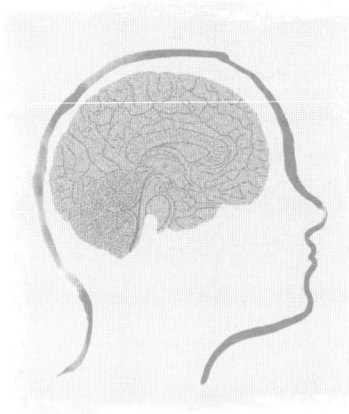

In recent years it has been found that *strategy training* for students with learning difficulties is one of the most promising areas of instructional intervention (Brennan & Robinson, 1998; Ellis, 2005; Graham & Bellert, 2005; Pressley, 1999). In particular, teaching students explicitly how to use comprehension strategies when interacting with text has produced impressive results (Swanson, 2000b). Strategies such as previewing, scanning, self-questioning, self-monitoring, rehearsing information and summarising have all proved valuable. One peer-assisted approach for teaching comprehension strategies is *reciprocal teaching* (RT) (see Chapter 4). Using RT, students can be taught to apply effective strategies for summarising, self-questioning, clarifying and predicting (Carter, 2001). Bruce and Robinson (2004) suggest that students with learning difficulties enjoy reciprocal teaching and are highly motivated. These researchers used RT in conjunction with a metacognitive training program to improve word identification for upper primary school students in Australia.

Reading comprehension skills can be increased when teachers spend time modelling and demonstrating effective strategies for processing text (Westwood, 2003a). Such strategies must encompass:

- Previewing material before it is read to gain an overview.
- Locating the main idea in a passage.
- Thinking aloud, and generating questions about the material.
- Predicting what may happen, or suggesting possible causes and effects.
- Summarising or paraphrasing main ideas.

A successful program for the development of comprehension should include at least these components:

- Large amounts of time devoted to reading.
- Using text as an important and valued source from which to obtain new information.
- Teacher-directed instruction in the use of comprehension strategies.
- Frequent occasions when students can talk with the teacher and with one another about their thoughts and responses to a particular text.

Reading strategies are best taught by direct means — through dialogue between teacher and students while working together to extract meaning from text. In this situation, dialogue allows students and teachers to become more aware of the process of making meaning from the printed page, and to learn from the successful strategies used by others. Dialogue also serves a diagnostic purpose by allowing the teacher to appraise students' existing strategies for comprehending (Westwood, 2003a).

Comprehension strategy training should always involve students using authentic texts for genuine purposes, and is therefore best taught within what are termed 'the content areas' of the curriculum, such as social studies, science, history and geography, as well as English. The use of contrived exercises for strategy training only in English lessons will show very little positive effect in terms of transfer and generalisation to other contexts.

The following general principles help to facilitate comprehension development for all students, including those with learning difficulties:

- Reading material should be interesting to the student and at an appropriate readability level.

- Students should always be aware of the purpose for reading a particular text.

- Prepare students for entry into a new text. Ask, 'What might we discover in this chapter?', 'What do the illustrations tell us?' or 'Let's read the subheadings before we begin'.

- If comprehension questions are to be set, encourage students to read and reflect upon them before the story or passage is read so that they enter the text knowing what information to seek.

- After reading a text, encourage students to set questions for each other, then use these questions to investigate the topic in more depth.

- Requiring students to make a summary is an excellent way of ensuring that they have identified main ideas.

- Graphic organisers can be used to summarise relationships among key points after reading the text.

- Use newspapers and magazine articles as the basis for some classroom discussion and comprehension activities. Highlighter pens can be used to focus upon key ideas, important terms or facts to remember.

Learning difficulties in writing

Written expression is often considered the most difficult of language skills for students to acquire. Competence in writing relies heavily on background knowledge and competence in listening, thinking, speaking and reading, as well as on possession of necessary strategies for planning, composing, encoding, reviewing and revising written language. For many years the teaching of writing, and the difficulties associated with it, received scant research attention compared to the volume of research conducted on reading and reading difficulties. However, that situation has changed significantly, and researchers and practitioners now contribute to the growing literature on teaching writing (e.g. Graham & Harris, 2000; Hess & Wheldall, 1999; Wong, 2000). In particular, increased interest has been shown in devising instruction that helps weak writers acquire some of the skills and strategies exhibited by proficient writers (e.g. Bereiter & Scardamalia, 1987; Graham et al., 1991).

Students with learning difficulties often exhibit weaknesses in planning, sequencing ideas, editing and revising their work (Topping et al., 2000). Some of these students also have problems with mechanical aspects of the task such as handwriting, spelling and punctuation. Gregg and Mather (2002, p.7) have remarked that:

> Writing competence is based on the successful orchestration of many abilities, including those needed for lower level transcription skills as well as those essential for higher level composing abilities … Students who struggle to develop written language often construct a negative perception of the writing process as well as a negative image of their own capabilities to communicate ideas through writing.

In other words, students who have problems writing will experience no joy or satisfaction in attempting the task, and will avoid writing whenever possible. Avoidance then leads to lack of practice, and lack of practice in turn results in no improvement. These students rapidly lose confidence and self-esteem in relation to writing, with both expository and narrative genres causing them many difficulties (Hammann & Stevens, 2003). Unfortunately, in school these forms of writing are frequently used as an integral part of the process of learning content material, and as a major form of assessment.

Approaches to the teaching of writing

There are three main approaches to teaching writing — the skills-based or

'traditional' approach, the 'process' or whole-language approach, and the 'socio-cognitive' approach. The skills-based approach provides explicit instruction, with guided practice and corrective feedback for writing in formal and informal styles. The approach is teacher-directed, with students developing writing skills mainly through structured exercises, practice materials (worksheets) and extended writing topics determined by the teacher.

When badly implemented, a skills-based approach can be boring for students and may have a detrimental influence on their attitude and motivation. Often the teacher using this approach simply sets a topic or exercise for students to tackle, but does not teach or advise them how best to go about the writing task. In this situation the activity becomes merely an assessment procedure rather than a teaching and learning experience (Pollington, Wilcox & Morrison, 2001). It is also believed that specific skills taught through exercises do not transfer to students' free writing or to writing in other subject areas. Instead of simply being given topics to write about, students need to be taught effective strategies for writing in different genres; when this is done, studies have shown that students improve in quantity and quality of written work and, equally importantly, adopt a better attitude toward tasks that involve writing (Hess & Wheldall, 1999).

Chan and Dally (2001) summarised possible benefits of teacher-centred instruction in writing, including the fact that some students clearly require explicit instruction in basic skills and strategies for writing because they do not learn them effectively through incidental exposure and immersion. Some students make better gains in writing through direct teaching, just as some students make optimum progress in reading if more time is spent in skill development through direct instruction and practice.

In the student-centred 'process' approach, students 'learn to write by writing', with much greater freedom to choose their own topics, and with less emphasis on formal teaching of grammar and style (Chan & Dally, 2001). An emphasis is placed on writing every day for genuine purposes of communication, and much less use is made of exercises that teach isolated aspects of the writing process, such as punctuation or paragraphing. Students are encouraged to experiment with ideas and express them in interesting ways. The written work is shared with the class and students receive feedback from teacher and peers that will help them revise and refine their written work. Process writing evolved mainly from the work of Donald Graves (1983) and has become fully integrated into whole-language philosophy.

Within the process approach, a method called *guided writing* can be incorporated (Riley & Reedy, 2000). Guided writing involves some degree of modelling by the teacher of specific strategies or genres, followed by guided application of the same principles and techniques by the students. Worthy, Broaddus and Ivey (2001) suggest that, using the overhead projector or computer, a teacher might demonstrate how to choose a starting point to begin writing, show several different types of leads to students and model the writing of a first paragraph. The teacher may also demonstrate in the same manner how he or she would go about editing, revising and improving the first draft of a story, description or report.

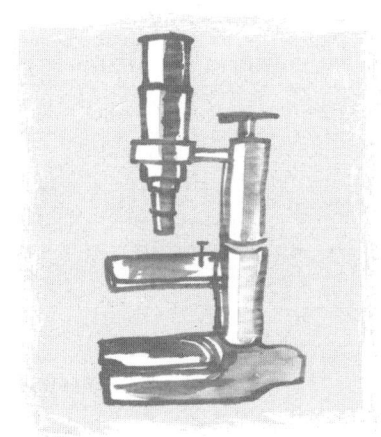

Later, students can take it in turns to present their own material to the group, and constructive suggestions and ideas can be shared in a type of reciprocal teaching situation.

The socio-cognitive approach can perhaps be seen as a combination of both teacher-directed instruction in writing strategies, and student-centred authentic writing. It is an approach that encourages extended writing, rather than the completion of routine worksheets and exercises. Hess and Wheldall (1999, p.15) state:

> Principles underlying the socio-cognitive approach are that literacy should be seen as having an important purpose by the children, that it is an interactive process through which children are supported as they learn new skills and put them into practice, and that the skills taught may be either broad or narrow depending on the writing needs of the children.

Hess and Wheldall (1999) indicate that helping students with writing difficulties involves using strategies to increase the amount they produce, to improve the clarity and organisation of their text, to engage students in collaborative efforts in composing and editing and to help them master writing in specific genres.

Strategy training

Students with learning difficulties appear to benefit greatly if they are taught strategies for generating and sequencing ideas, composing and revising (Harris, Graham & Mason, 2003; Wong et al., 1996). They need to be given guidance in how to begin, how to continue and how to complete a writing task. Bereiter and Scardamalia (1987) suggest using the following set of prompts to help students:

1 Identify the goals for writing.
2 Generate a new idea.
3 Elaborate or extend an idea.
4 Sequence ideas into a cohesive whole.

The best starting point for strategy training is explicit modelling by the teacher of an effective writing strategy suitable for the age and ability level of the students. The teacher uses a whiteboard or the computer screen while 'thinking aloud' to reveal the way he or she goes about each step in planning, generating ideas, drafting, revising and polishing a piece of writing. Students then practise the application of the same steps in the process and receive feedback from the teacher.

Word processors

Word processors have opened up many new opportunities to motivate poor or reluctant writers, as well as providing all writers with an improved system for composing and editing. Phillips (2004) identified several advantages in using word processors to encourage students in their writing, including the following:

- Mistakes can be corrected without a mess.
- When printed, the finished product looks very professional.
- Revisions are made more easily.
- Tedious rewriting after editing is not required.
- Students can work collaboratively in the composing and revising stages.
- Finished products can be shared and critiqued easily within the peer group.

Teaching spelling

While it is true that there are a few individuals who appear to have a natural aptitude for accurate spelling and therefore require very little guidance, the majority of learners are not so blessed and do need teaching if they are to become good spellers. Research suggests that weak spellers can become more proficient if appropriate instruction is provided for them (Graham, 2000; McNaughton, Hughes & Clark, 1994).

Skill in spelling relies on the effective use of eyes, ears, voice, brain and hand (Westwood, 2005). Eyes are involved in perceiving and storing accurate images of orthographic patterns, and in detecting whether a word looks correct after it has been written. Ears are involved in detecting the sequence of sounds within a spoken word that provide some clue to the way in which the word may be written. Voice is involved to the extent that articulating a word as a supplement to hearing the word can provide clues to its possible spelling. Brain is involved in relating sounds to letters or groups of letters (phonic information) and in connecting the meaning of a word or a part of a word with its spelling. The hand is involved in that skilful spelling requires an automated and swift motor response. These motor responses become established through frequent practice in writing a word. The teaching of spelling needs to take into account these several contributory factors.

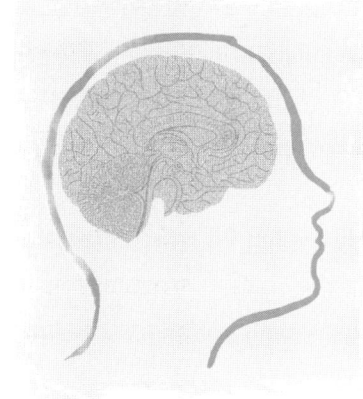

The main methods for teaching spelling are:

- *The whole-word visual approach:* This method is essential for learning to spell irregular words that do not lend themselves to phonic translation (e.g. *ache, work, cough, choir, answer*). The strategy that capitalises on this visual method is referred to as Look-Cover-Write-Check (Peters, 1985), sometimes modified to Look-Say-Cover-Write-Check to include the influence of pronouncing a word while studying it to help establish its spelling pattern.

The aim of this method is to strengthen visual imagery of word forms. There is some evidence that the 'concreteness' of a word (how meaningful it is to the speller, and how much it represents real objects or experiences in the speller's mind) is an aid to its storage in visual form (Sadoski et al., 2005).

- *The phonemic (or phonic) approach:* In this method the speller is encouraged to sound out a word and write it using phonic principles. The teaching usually progresses from spelling simple consonant-vowel-consonant words to mastery of more complex letter groups representing pronounceable parts of words (e.g. *-eed, anti-, dis-, -ure, -each*).

- *The morphemic approach:* A morpheme is the smallest unit of meaning within a word. A word may comprise one morpheme (e.g. *sane* = 1) or more (e.g. *insane* = 2, *insanity* = 3). The word *insanity* illustrates that under the morphemic approach the learner is also taught to apply certain rules when changing a root word (*insane* drops the *e* when *ity* is added). Direct instructional methods are usually employed for a morphemic approach, and the program *Spelling through morphographs* (Dixon & Engelmann, 1976) is a good example.

- *Spelling by analogy:* Students are taught to reflect upon words they can already spell correctly to help them work out the probable spelling of an unfamiliar word. For example, knowing how to spell the word *each* helps one to spell *teach, teacher* and *teaching*. Using visual checking should indicate to the speller that *teeching* looks wrong.

- *The multisensory approach:* This method, involving finger-tracing over a written version of the word while looking at it and saying it clearly, is slow and tedious but often helps students with specific learning disability (and students with intellectual disability) to store a visual and motor image of the word in long-term memory. With other students it can be used as a last resort to overcome a particular spelling demon.

  - *The strategic approach:* If students are to become truly independent in their spelling they need to be able to determine for themselves how best to spell a new word. Is it a regular word that can be written directly by translating sounds to letters? Is it an irregular word needing a visual approach, perhaps supplemented by repeated writing? Can I spell this word by comparing it mentally with a word I already know? What is the most difficult and unpredictable part of this word? A general purpose spelling strategy is illustrated on the following page (Lam & Westwood, 2006):

I must ask myself:

— Do I know this word?

— How many syllables can I hear when I say the word?

— Do I know any other word that sounds almost the same?

— Which letter groups do I need to write?

— Does the word I have written look correct?

— I'll try again.

— Does this look better? Let me check.

The teaching of spelling should not be confined to the realm of the English teacher. All subject teachers should accept the responsibility of encouraging accurate spelling in their own area of the curriculum. It is valuable if a subject teacher compiles a core vocabulary list for his or her subject and makes this available to all students in the form of a self-help spelling aid for use when written work is undertaken in that subject.

Useful resources

Teaching students to read

Westwood, P. (2001). *Reading and learning difficulties.* Melbourne: Australian Council for Educational Research.
Contains more detailed information on learning difficulties in the literacy domain.

Phonics skills

Learning and Teaching Scotland (2005). *Synthetic v analytic phonics.* Dundee: Learning and Teaching Scotland. Viewed 2 February 2006, (http://www.ltscotland.org.uk/5to14/specialfocus/earlyintervention/issues /phonics.asp).
Useful information comparing synthetic and analytic phonics approaches.

Comprehension

Barr, R., Blachowicz, C., Katz, C. & Kaufman, B. (2002). *Reading diagnosis for teachers* (4th edn). Boston: Allyn & Bacon.
Dymock, S. & Nicholson, T. (1999). *Reading comprehension: What is it?* Wellington: New Zealand Council for Educational Research.
Additional information on improving reading comprehension in students with learning difficulties.

Spelling

Westwood, P. (2005). *Spelling: Approaches to teaching and assessment* (2nd edn). Melbourne: Australian Council for Educational Research.
Contains information on general and remedial approaches to spelling instruction.

General

Carnine, D.W. (2004). *Direct instruction reading* (4th edn). Upper Saddle River, NJ: Pearson-Prentice Hall.

Cooper, J.D. & Kiger, N. (2006). *Literacy: Helping children construct meaning* (6th edn). Boston: Houghton Mifflin.

Lerner, J. & Kline, F. (2006). *Learning disabilities and related disorders: Characteristics and teaching strategies* (10th edn). Boston: Houghton Mifflin.

Tompkins, G.E. (2006). *Literacy for the 21st century: A balanced approach* (4th edn). Upper Saddle River, NJ: Pearson-Merrill-Prentice Hall.

These texts are also pertinent to the issues raised in this chapter.

6 Teaching basic academic skills: mathematics

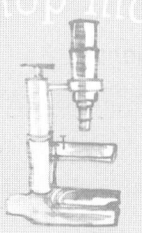

Students with learning problems frequently have difficulty acquiring and maintaining basic mathematics computation and problem solving skills. Although many students with math deficiencies exhibit characteristics that predispose them to math disabilities (e.g. problems in memory, language, reading, reasoning and metacognition) their learning difficulties are often compounded by ineffective instruction (Miller, 1999, p.165).

The educational psychologist Ormrod (2000) points out that mathematics probably causes more confusion and frustration, for more students, than any other subject in the curriculum. In particular, the sequential and hierarchical nature of mathematical learning means that a student who fails to master early concepts and skills is destined to have increasing difficulties as new and more demanding material is introduced in a course. Students with learning difficulties often give up all hope of ever understanding what mathematics is about, and have resorted to rote memorisation of meaningless facts, symbols and procedures. Not surprisingly, the majority of such students also develop a very negative view of mathematics as a subject, and of their own competence with number.

In Australia, the writers of the *National statement on mathematics for Australian schools* (Curriculum Corporation (Australia), 1991, p.7) remarked:

> There is considerable anecdotal and research evidence to suggest that many people dislike mathematics and may even feel intimidated in situations in which it is used. Of considerable concern is the effect on individuals of having to deal with an increasingly mathematically oriented society while feeling inadequate or alienated from mathematics.

In this chapter attention will be given to the possible reasons why so many students find this subject unfathomable. How should mathematics be taught to make it more accessible to a wider range of learners?

Goals and content of mathematics education

The importance of teaching mathematics to all students is captured well in the *National Curriculum* guidelines published in the UK (QCA, 2005, p.1):

> Mathematics equips pupils with a uniquely powerful set of tools to understand and change the world. These tools include logical reasoning,

problem-solving skills, and the ability to think in abstract ways. Mathematics is important in everyday life, many forms of employment, science and technology, medicine, the economy, the environment and development, and in public decision-making.

The curriculum framework document prepared by the Curriculum Council in Western Australia adds:

> In mathematics, students learn to use ideas about number, space, measurement and chance, and mathematical ways of representing patterns and relationships, to describe, interpret and reason about their social and physical world. Mathematics plays a key role in the development of students' numeracy and assists learning across the curriculum (Curriculum Council (Western Australia), 1998, p.1).

In most countries, mathematics education tends to be based on the following general goals designed to foster students' mathematical literacy (Harniss et al., 2002):

- Students will develop a positive attitude toward mathematics and will recognise its value.
- They will become confident in their ability to understand and use mathematics.
- They will develop effective problem-solving skills and strategies, and will learn to reason mathematically.
- They will be able to communicate mathematically.

Official curriculum guidelines in the UK and Australia add another important goal — that the learning of mathematics should bring with it enjoyment and satisfaction, in addition to building functional skills and understandings.

The main curriculum documents specifying goals and content and influencing mathematics teaching in Australia continue to be the *National statement on mathematics for Australian schools* (Curriculum Corporation (Australia), 1991) and the companion volume *Mathematics: A curriculum profile for Australian schools* (Curriculum Corporation (Australia), 1994a). In Britain, mathematics content and outcomes are specified clearly within the *National Curriculum* (see below). General goals for teaching mathematics are usually translated into more practical terms by setting them out as objectives or standards for each domain of study and for each age level throughout school. Examples of such objectives can be located through resources listed at the end of this chapter.

In almost all countries, the curriculum in mathematics typically covers the following domains of study (although the areas may be differently named in some countries): number and number operations, measurement, algebra, geometry, data analysis and probability, and problem-solving. In Britain, the four broad areas for which learning targets are specified in the *National Curriculum* include: using and applying mathematics; number and algebra; shape; space and measurement; and handling data. The *National Curriculum* sets out clearly the attainment targets for knowledge, skills and understandings that students of different ages and abilities are expected to have by the end of each of four 'key stages'. Broadly speaking, Key Stage 1 is from beginning school to age seven years; Stage 2, between seven and 11 years; Stage 3, from 11 to 14 years; Stage 4, from age 14 to the end of compulsory schooling. In mathematics, Stage 4 is further subdivided into 'Foundation' and 'Higher achievement'. Attainment targets within each stage consist of eight 'level' descriptions of increasing difficulty. Level descriptions indicate the competencies that a student is expected to demonstrate at each level. Examples of these competencies can be located through resources listed at the end of this chapter. Students take national competency tests in mathematics at the end of each key stage. For students with significant learning difficulties or a disability, a level 'P' has been created, identifying the knowledge and skills required for readiness to enter Level 1. This equates with the 'Towards Level 1' category used in the Australian profiles for mathematics.

Teaching and learning in basic mathematics

Traditionally, the teaching of basic mathematical skills involved drilling children in routine arithmetic processes, leading ultimately to mastery of the algorithms required for addition, subtraction, division and multiplication, together with some (often very abstract) work with fractions. Almost by way of relief from drudgery, children might be exposed to small amounts of geometry, measurement and problem-solving. When this writer began teaching in primary schools in England in the 1950s, emphasis was almost entirely on teaching computational skills, with books such as Lovell and Smith's (1956) *Two-Grade Arithmetic* and Schonell's (1949) diagnostic number tests very much in evidence in most classrooms. Gradually, in the 1960s, the typical arithmetic syllabus began to include more interesting topics, influenced to some degree by the theories of Piaget on children's conceptual development in number and the need for active learning (Isaacs, 1960). This, together with the impact of the Nuffield Primary Mathematics Project and a new textbook series by 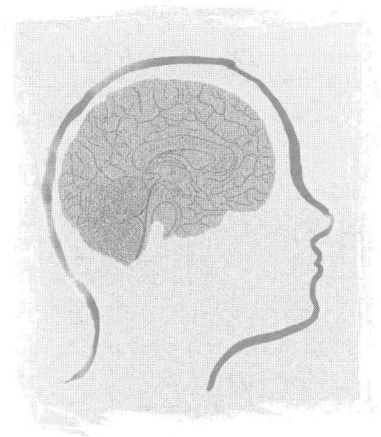 Flavell and Wakelam (1960), led gradually to a reduction in the amount of abstract and decontextualised calculation in the primary curriculum. Teachers were encouraged to provide more activity-based learning involving group work, inquiry and discussion. At the same time, great enthusiasm was being shown for the use of 'structural apparatus' and manipulatives, such as Cuisenaire rods (Cuisenaire & Gattegno, 1954; Gattegno, 1960) and *Colour factor mathematics* (Thompson, 1962) to help students 'discover' number relationships through concrete, hands-on activities. These more student-centred approaches continued to develop into the 1990s, supported fully in most developed countries by the official guidelines for teaching mathematics. Student-centred and problem-based approaches became the focus of initial teacher education courses and in-service programs for teachers of mathematics. The situation remains largely the same now.

Contemporary approaches

Current reforms in mathematics education are based on socio-constructivist theories of learning, and place much emphasis on practical problem-solving and investigation. Even with students of limited ability, the application of number skills to solve real-life problems is advocated as the desirable approach to use, rather than simply drilling arithmetic processes through practice exercises (Parmar, Cawley & Frazita, 1996). It is argued that mechanical arithmetic skills are worthless if they cannot be applied automatically and effectively to situations where they are needed. The NCTM (2005) states that problem-solving is not only the *goal* of mathematics education, but is also the *process* by which meaningful learning takes place.

The current views of mathematics teachers who subscribe to constructivist principles are that mathematics should be taught using a predominantly inquiry approach which embraces elements of discovery, investigation, cooperation, situated learning, project work and cognitive apprenticeship (see Chapter 3). Boaler (1997) suggested that contemporary approaches to mathematics learning must become much more 'authentic' and less 'algorithmic' in order to produce knowledge and skills that are adaptable to new contexts and can be applied in the real world. This view remains dominant in almost all curriculum guidelines published in recent years.

Advocates for what is now termed 'process math' (or 'whole math') take much the same stance as taken by whole-language exponents in the literacy domain.

They believe that facts and skills must be acquired *during* the process of investigating authentic quantitative situations, not as a preparation for engaging in such activities later (Wakefield, 1997). They favour an orientation toward discovery learning rather than expository teaching, and they believe that cooperative group work is much more productive than any teacher-directed lesson. The most extreme advocates for activity-based mathematics shun all practice exercises designed to build up computational skills, and abhor the rote memorisation of multiplication tables. Instead, they welcome students' use of unconventional methods for obtaining solutions, and recommend that students use calculators rather than spending time and effort trying to commit number facts to memory.

Is the pendulum swinging again?

By the end of the 1990s, some elements of constructivist approaches to the teaching of basic mathematics were coming in for criticism from some quarters. The notion that mathematics can be taught successfully using the same type of unstructured student-centred immersion approach advocated for literacy learning has been challenged (e.g. Hunting, 1996; Jones & Southern, 2003; Klein et al., 2005). For example, it has been argued that calculators are now used too much in the primary years, defeating the aim of having students acquire 'number sense' and understand the processes used in arithmetic. In the early years, manipulatives and concrete aids are perhaps used too much and for too long, so that students become dependent on them rather than developing mental skills and strategies for dealing with numbers. It is suggested that the pendulum has perhaps swung too far away from teaching arithmetic skills so that too many students lack automaticity in dealing with number facts (Westwood, 2003b). Some students cannot perform algorithms accurately or with confidence, and many have a poor understanding of fractions (Klein et al., 2005).

Added to these problems, there is a general belief that students with learning difficulties do not do well in unstructured inquiry-type mathematics programs, but achieve far more when taught by direct methods (Ellis, 2005). There is much support in the research literature for the use of Direct Instruction (DI) in the development of both number skills and problem-solving abilities (Farkota, 2005; Pincott, 2004; Silbert, Carnine & Stein, 1990); but outside the context of remedial teaching, DI is not widely used for basic mathematics, mainly because it appears to go against the flow of an 'inquiry and problem-solving approach'. Typical DI lessons have much to offer in establishing confidence and automaticity. In particular they hold students' attention very well, and they ensure a  great deal of daily practice and application (Serna & Patton, 1997b). The relative merits and problems associated with DI were discussed in Chapter 2, and Ellis (2005) has reviewed research studies involving DI in numeracy teaching.

It is not surprising that there has been a call from some educators to redress the balance in mathematics teaching so that all students acquire fundamental number skills (Jones & Southern, 2003). As the NCTM (2005) points out, computational skills and number concepts are essential components of the mathematics curriculum, and the ability to estimate and carry out mental computation is more important than ever. Similarly, McIntosh and Dole (2000) indicate that mental computation and number sense need to become integrated components of mathematics lessons again, to avoid the curriculum becoming distorted by undue emphasis on problem-solving. Procedural fluency with numbers, strategic competence and adaptive reasoning in problem-solving all remain vital goals in mathematics education (Kilpatrick, Swafford & Findell, 2001), but their mastery cannot be guaranteed by student-centred inquiry methods alone.

The conclusion that one must reach is that, just as in the case of literacy, a balanced approach to the teaching of mathematics is necessary, and must include a significant measure of explicit teaching together with valuable hands-on activities, exploration and problem-solving that typify constructivist programs (Burns, 1998). The amount of explicit instruction required in teaching mathematics varies from topic to topic and from student group to student group, with abundant direct teaching often being of most benefit for students with learning difficulties.

Effective teaching of basic mathematics

In the 1980s, much of the research into effective teaching used mathematics lessons to collect data. This was partly because students' achievement in mathematics is more easily measured objectively than is the case with achievement in many other subjects in the curriculum. As a result of the studies cited in Chapter 2, and of the more recent work on teaching methods in mathematics (e.g. Brown, 2000; Reynolds & Farrell, 1996; Stigler & Hiebert, 1999), it is now believed that effective lessons appear to incorporate the following features:

- Interactive whole-class teaching, with high levels of student participation.
- Direct teaching of basic skills, with sufficient practice to establish these skills to a highly functional level.
- Frequently asking students questions to stimulate thinking and to check for understanding.
- Adequate attention to number facts, mental arithmetic and problem-solving.
- Appropriate pacing of the curriculum.
- Introducing concepts that are developmentally appropriate for the age group.
- A teacher who helps students to establish meaningful connections between concepts.
- Use of visual and concrete materials to provide a starting point for establishing concepts and skills.
- Frequent use of authentic materials and real-life problems.
- Effective use of the blackboard (or interactive whiteboard) to summarise steps involved as a problem is solved.

- Explicit teaching of strategies for analysing and solving numerical problems.

- Close monitoring of students' work by the teacher, resulting in the opportunity to reteach and provide corrective feedback.

- Balancing appropriate use of student-centred group work, discussion, individual work and teacher-led expository instruction.

- Effective use of peer assistance and peer tutoring when appropriate.

- Use of ICT to provide simulations for situated learning, computing aids and access to data.

- Regular assessment of learning to indicate effectiveness of the program.

- Frequent revision of previously taught concepts and skills.

- Differentiating homework tasks according to students' learning needs.

There are some indications that students develop numeracy skills most effectively in classrooms where teachers have an academic focus, use challenging activities, have high expectations of students (including lower achievers) and do not rely entirely on whole-class teaching or entirely on discovery learning and individual progression (Brown, 2000). The studies of teaching in countries with high achievement data in international surveys (e.g. Stigler & Hiebert, 1999) show that the teachers tend to devote much more time to encouraging students to think and reason while problem-solving, and avoid routine application of algorithms. Discussion, and critical thinking about possible solutions and alternative ways of tackling problems, forms a large part of each lesson. Students' ideas are recorded on the whiteboard and analysed for their suitability in the given context. Ormrod (2000) comments that a growing body of research evidence supports the effectiveness of group discussions in enhancing students' mathematical understanding through the sharing of ideas. In general, countries such as Japan, Singapore and Korea require students to adopt a much deeper learning approach in mathematics than is typical in most Western countries.

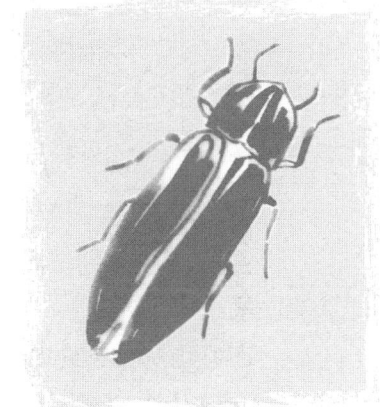

Poor-quality teaching

By way of contrast, poor-quality teaching in mathematics tends to be identified by:

- An entirely textbook-driven approach that does not link with students' lives or interests.

- Introduction of curriculum content that requires cognitive abilities beyond those the students possess at that stage.

- New topics taught in isolation instead of linking and relating topics conceptually.

- Too much content covered in the curriculum, leading to superficial learning.

- Too much teacher talk (lecturing) with too little discussion and questioning.

- Inability of the teacher to explain or demonstrate at a level that students can understand.

- Lack of visual and concrete materials to support new concepts and to stimulate interest.

- Unstructured discovery approach, leading to confusion.

- Insufficient modelling by teacher of effective methods for calculating and recording.

- No variety provided within the lessons.

- Too much attention devoted to mechanical arithmetic with too little application to real problems.

- Too little attention to mechanical arithmetic, leading to lack of speed and accuracy when solving problems.

- Too much time spent in group work.

- Too little time spent in group work.

- Students always being forced to work individually and not cooperatively.

- Inadequate revision of previously taught concepts and skills.

- Inadequate monitoring of individual progress.

- Lack of assistance for students who are experiencing difficulties.

Difficulties associated with learning basic mathematics

The quotation at the beginning of this chapter points out that the quality of teaching is a major influence on whether or not students experience difficulties when learning mathematics. Booker (2004, p.130) remarks that, 'It needs to be acknowledged that many students have *learned difficulties*, rather than learning difficulties [in mathematics]'. He suggests that teachers themselves often have only a 'pseudo-conceptual' understanding of mathematics, rather than a deep understanding of the subject matter. This lack of what Shulman (1987) referred to as essential subject knowledge and pedagogical content knowledge (see Chapter 1) leads them to teach the subject in a rather disconnected and abstract procedural manner. They may fail, for example, to teach students key relationships connecting division, fractions and ratios, but instead teach each concept quite separately. They may, as Booker indicates, teach operation rules without intrinsic meaning (e.g. 'carry one', 'move the decimal place to the left', 'always put down a zero first'). One of the clearest indications of 'good' teaching in mathematics is whether the teacher helps students see connections within the subject, and does not move on to new work until understanding is established (Brown, 2000).

Meaningful learning in mathematics requires that such connections be made (Serna & Patton, 1997b).

Several writers have described difficulties that students encounter when learning mathematics (e.g. Booker, 2004; Lamb, 2004; Pincott, 2004). These difficulties include:

- Inaccurate counting.
- Poor recall of number facts.
- Confusion over steps with an algorithm.
- Inability to determine the appropriate processes to use to perform a calculation and to solve a problem.
- Lack of understanding of place value.
- General slowness in completing calculations, resulting in less practice and failure to establish automaticity with even the simplest steps in a calculation.
- Signs of possible limitations of working memory capacity.
- Untidy bookwork and recording, leading to careless errors and confusion.
- Difficulty with reading and understanding problems in word form.
- Having no appropriate strategies for approaching problems.
- Lack of number sense, or an ability to estimate, thus being unable to detect when a numerical solution is obviously incorrect.
- Very poor transfer and generalisation of skills taught in mathematics lessons to other relevant contexts.

Students may have specific areas of weakness even though their overall number skills are adequate. For example, Lamb (2004) reports that students have particular difficulty with the division process. This may be because division is usually taught long after students have been weaned away from concrete materials. In such cases, it is taught as a rote algorithmic procedure with paper and pencil, not by moving (sharing) real items from a large group into smaller groups. Without such practical experience the students have no clear evidence of what the algorithm represents. This leads in turn to problems of not understanding what a 'remainder' is, not being sure how to perform the division algorithm starting on the left (because all other processes begin on the right), and failing to establish any connection between division, fractions and ratios. It is probable that division and fractions  are the most inefficiently taught concepts in primary mathematics; certainly they account for many of the errors that students make. Because students may have weaknesses in certain processes it is often recommended that teachers study the errors students make in their written work in order to identify any misconceptions (Ashlock, 1998; Booker, 2004).

When we probe beyond the quality of instruction to investigate learning difficulties in mathematics we may find the following problems:

- *Poor attention to task:* In previous chapters, the importance of attention has been stressed. While inattention can often be due to frustration and boredom when lesson content is not making sense, it can also be due to such factors as fatigue, poor health, anxiety or attention deficit disorder. As in all teaching situations, the onus is on the teacher to establish good attending behaviour among members of a class.

- *Learned helplessness:* If a student experiences almost constant failure in mathematics lessons, it is not surprising that belief in their own ability is diminished and they fall into a cycle of failure (Brophy, 1998). Interest, motivation and self-esteem suffer, leading to passivity.

- *Anxiety:* It is well documented that mathematics can cause high levels of anxiety in some students. Such anxiety is detrimental to confidence and attention (Tobias, 1993).

- *Memory:* Miller (1999) cites problems with memory as one possible contributory cause for learning difficulties; the research of Hopkins (1998) suggests that capacity deficits in working memory may be implicated in students' difficulty in rapid recall of number facts. It is true that some individuals do have more difficulty than others in storing and retrieving information, but poor recall is usually due to some combination of failing to pay attention, failing to practise (rehearse) information, or information that was of no interest or did not link in any way with previous learning.

- *Specific learning disability:* In the same way that a small percentage of students have chronic difficulties learning to read, write and spell despite having at least normal mental ability (dyslexia), there is believed to be a small number of students with a similar problem in dealing with mathematics (dyscalculia) (Temple, 2001). Students with dyscalculia are said to exhibit lack of understanding of mathematical terms and symbols, weak number sense, poor accuracy in applying algorithms (particularly multiplication and division), untidy book work, reversal of individual digits and of place value, inability to detect unfeasible answers and a lack of flexibility in their approach to problem-solving. These students usually require intensive one-to-one tuition if they are to make any progress; even then, their personal and emotional reactions to their difficulties make such tuition an uphill battle. For more information on dyscalculia, see Westwood (2004).

Strategy training for problem-solving

Chapter 3 emphasised the value of teaching students how to approach particular tasks with a plan of action in mind. In the domain of mathematical problem-solving, cognitive strategy training of this type has proved to be very effective in improving students' performance (e.g. Montague & Bos, 1986; Owen & Fuchs, 2002).

Strategy training for mathematics usually involves teaching students a 'self-talk' script to help them regulate their approach to the problem. The method of instruction might be termed 'teaching problem-solving as a thinking process'. First the students are taught to ask themselves what the problem requires them to find out. Then they determine which information contained within the problem is needed for calculating a solution, and they decide which arithmetic process (or processes) will be used. They complete the calculation and then ask themselves, 'Does my solution make sense? Have I taken into account all the relevant information?' They double-check the solution and confirm their calculations, or self-correct if necessary.

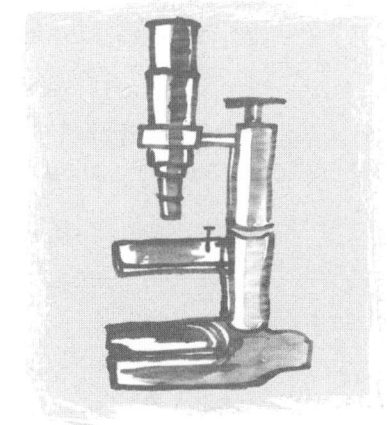

Consider the following problem: 253 soldiers are waiting to be moved from one army camp to another, a journey of 37 kilometres. They will be transported on trucks that can each carry 40 soldiers (and no more). How many trucks will be needed to complete the transfer of soldiers if trucks can return and do a second (but not a third) journey?

Students ask themselves:

- What does the problem ask me to calculate? (The number of trucks needed.)
- Which are the important numbers to use? (253 soldiers; 40 per truck.)
- Is there other information I must use? (Yes: the soldiers are not all moved at the same time; the trucks do a return journey.)
- Are there numbers and details that I can ignore? (Yes: 37 kilometres is not relevant.)
- Will it help me if I draw a sketch? (Yes, it may help, but I don't need to draw 253 soldiers. I do need to draw lines showing trucks going between the camps twice, and I can write '40' against each line.)
- What process will I need to use with the 253 and 40? (Divide: $253 \div 40$.)
- That is easy, I can work it in my head, but I will write it down anyway. ($253 \div 40 = 6$ remainder 13.)
- 6 what? (Trucks.)
- Does that make sense? (No, I can't have 6 trucks, remainder 13.)
- What do I do with the remainder? (Do I divide 13 by 40 to express it as a decimal fraction? No, that does not make sense. I can't have a fraction of a truck.)
- What does the remainder represent? (13 soldiers who can't be carried on the trucks on either journey.)

- OK, so I need another truck to take the 13 soldiers. (So 7 trucks are needed.)
- So, is that the solution? Is it 7 trucks? (No: the trucks do a double journey, so I only need 3.5 trucks. No, wait. I can't have half a truck. I need 4 trucks, but 1 will only do a single journey.)
- Let me check this. I will write it down:
  — First journey: 40 + 40 + 40 + 13 = 133. Second journey: 40 + 40 + 40 = 120.
  — 133 + 120 = 253.
  — So 3 trucks do a double journey with 40 soldiers each time; 1 truck does a single journey with 13 soldiers.
  — Correct: 4 trucks would be needed.

When teaching problem-solving strategies of this type, the teacher normally models the process clearly, using the whiteboard to record steps and procedures while 'thinking aloud' (Miller, 1999). The thinking involved in applying a problem-solving strategy is illustrated in the example above. In the interactive whole-class teaching sessions (typical, for example, in Japanese classrooms) the process of solving the problem is modelled and analysed at each step. Students are asked to explain or justify the thinking behind each suggestion they may make. The intention is to equip students with a logical approach that relies not upon rote application of arithmetic skills to routine problems, but upon reasoning, understanding and self-monitoring. When errors are made, students have the confidence to self-correct and to check again for feasibility and accuracy.

When assessing students' problem-solving skills and strategies, it is usually necessary to spend time with each individual, and to ask them to explain and demonstrate their thinking. In particular, it is important to observe the extent to which they can complete some of the steps mentally and to which they require semi-concrete aids to help them visualise or understand a problem (e.g. sketches, tally marks, counters).

Using appropriate methods to match instructional aims

In summary, basic mathematics can be taught at all age and ability levels using a wide range of methods, from teacher-directed expository instruction to student-centred activity and exploration. Research has tended to support the view that direct teaching is necessary when introducing and establishing basic skills for calculation and estimation, especially when working with students with learning difficulties (e.g. Butler et al., 2001; Tournaki, 2003). Such teaching is usually accompanied by abundant opportunities to practise these skills to mastery level. For students who may be struggling to achieve competence, approaches such as precision teaching and mastery learning have much to offer (see Chapter 2). However, since the goal of teaching mathematics is to equip individuals with skills and strategies for solving real problems and for dealing with the quantitative aspects of everyday life, student-centred methods involving investigation and the application of appropriate strategies must also be key components in any mathematics curriculum. The cognitive apprenticeship approach is particularly appropriate for teaching problem-solving strategies.

Useful resources

General

NCTM (National Council of Teachers of Mathematics) (2005). Reston, VA: NCTM. Viewed 2 February 2006, (http://www.nctm.org).
Teachers will find much of interest on this website.

Curriculum — US

NCTM (National Council of Teachers of Mathematics) (2005). *Number and Operations Standard.* Reston, VA: NCTM. Viewed 2 February 2006, (http://standards.nctm.org/document/appendix/numb.htm).
Contains examples of mathematics objectives, together with the related curriculum content.

Curriculum — UK

DfES (Department for Education and Skills) (2006). *Primary national strategy: Framework for teaching mathematics.* London: DfES. Viewed 2 February 2006, (http://www.standards.dfes.gov.uk/primary/publications/ mathematics/math_framework).
This online document illustrates the intended range of work for numeracy development. Its purpose is to help teachers in primary, middle, and special schools set appropriate expectations for their students.

QCA (Qualifications and Curriculum Authority) (2001). *Mathematics: Introduction.* London: QCA. Viewed 2 February 2006, (http://www.ncaction.org.uk/subjects/maths/index.htm).
For general information on mathematics in Britain's National Curriculum.

QCA (Qualifications and Curriculum Authority) (2001). *Mathematics: The level descriptions.* London: QCA. Viewed 2 February 2006, (http://www.ncaction.org.uk/subjects/maths/levels.htm).
Level descriptions for mathematical competencies at each age level within the National Curriculum in Britain.

Special education

DfES (Department for Education and Skills) (2006). *Key Stage 3 National Strategy: Accessing the National Curriculum for Mathematics.* London: DfES. Viewed 2 February 2006, (http://www.standards.dfes.gov.uk/keystage3/ respub/ma_access_nc).
Provides additional information on what students with special education needs might be expected to achieve in mathematics.

Teaching methods

Carnine, D., Dixon, R. & Silbert, J. (1998). Effective strategies for teaching mathematics. In E. Kameenui & D. Carnine (eds) *Effective teaching strategies that accommodate diverse learners* (pp.93–112). Columbus, OH: Merrill.

Westwood, P. (2000). *Numeracy and learning difficulties.* Melbourne: Australian Council for Educational Research.

Effective teaching methods for mathematics are described more fully in these two texts.

7 Teaching science

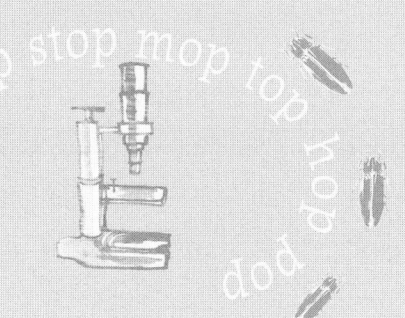

Science is diverse and exciting. It helps pupils explore the world around them and understand so many things that have such relevance to daily life. Pupils must, therefore, have the best possible support for learning science at school (DfES, 2002a, p.3).

In most education systems worldwide, helping students become scientifically literate is regarded as one of the important goals of the school curriculum. Scientific literacy has been defined as 'the capacity to use scientific knowledge, to identify questions and to draw evidence-based conclusions in order to understand and help make decisions about the natural world and the changes made to it through human activity' (OECD, 2000, p.76).

Perhaps the most universally shared understanding among science educators and policy-makers today is that the study of science is for *all* students, not just for those in secondary schools who have high ability or special aptitude. This theme is evident, for example, in the documents *Science for All Americans* (AAAS, 1990) and *Benchmarks for Science Literacy* (AAAS, 1993) published by the American Association for the Advancement of Science, and in the several supplementary guidelines provided to accompany the *National Curriculum Frameworks* in Britain (e.g. DfES, 2002a; NCC, 1992). Students with learning difficulties or disabilities are acknowledged to have a full entitlement to engage in science activities along with their peers, and to be provided with whatever support may be required to help them participate and learn in regular classrooms or in special schools (Fenton, 2002).

Science for all

The notion of 'science for all' has particular importance for students with special educational needs situated in mainstream inclusive classrooms. Naturally, these students will share in the same learning experiences as all other students and will participate in the same activities (or in adapted versions of the same activities) to the best of their ability. For some students with intellectual, physical or sensory disabilities, the science activities, resources and teaching methods may need to be modified quite considerably to accommodate their learning characteristics and special needs (ASE, 1997). Any adaptations that are made to the ways in which science is taught to these students should still respect the overall goals for science education as applied to all students. These goals typically embrace beliefs that:

- All students should experience the richness and excitement of knowing about and understanding the natural and constructed world.

- All students should understand and use appropriate scientific skills and processes, such as observation, investigation, experimentation, measurement and data collection, prediction, comparison, discussion, verification and decision-making.

- All students should be helped to develop positive attitudes toward science as a subject, and to value the contribution science has made and will continue to make to human achievement, to medical and technological progress and to quality of life. Attitudes and values to be developed include curiosity, wonder, respect and personal responsibility.

In order to achieve these broad goals, students will need to engage in activities that help them acquire accurate factual information, develop concepts, understand cause and effect relationships, and recognise applications of scientific principles in everyday life. Students with learning difficulties may require considerable support in working toward these goals.

Approaches to teaching and learning

There are many excellent ways of providing positive learning experiences for students in science, and appropriate teaching methods can range from a student-centred discovery approach through to direct instruction (Weiss et al., 2003). There is an unfortunate impression that all science lessons should be entirely activity-based discovery, with little or no direct input, explanation or instruction from the teacher. Perhaps this impression is fostered by the professional literature on constructivist approaches to learning, and by the recommendations for methodology given in curriculum guidelines. The reality is that almost all the teaching and learning methods described in Chapter 2 and Chapter 3 can be used or adapted for the teaching of science. Science experiences can be provided via explicit instruction; interactive whole-class teaching; cooperative group work involving activity-based, task-based or problem-based learning; project work; peer tutoring; individual contracts; and computer-assisted learning. Frequent use can also be made of laboratory work, field trips, video resources, computer simulations, models, textbooks and discussion.

The methods selected for use in any particular lesson should match as far as possible the cognitive level and learning characteristics of the group of students being taught, and should also match the *type of learning* involved in achieving the lesson objectives. For example, the teaching processes required to establish firmly the students' *declarative knowledge* (covering scientific facts, terminology, rules and concepts) may be different from the type of instruction needed to establish *procedural knowledge* (knowledge of effective ways for carrying out systematic observations, assembling equipment, conducting investigations and processing data). Similarly, a science lesson with a focus on influencing beliefs and attitudes concerning the value of science to the community (*affective learning*) may very well require quite different activities from one that requires students to discover a principle or deepen their understanding of a concept (*conceptual learning*).

Virtually all official curriculum guidelines issued in the past 20 years stress the value of the 'hands-on' investigative aspects of good science teaching, and emphasise the 'facilitator' rather than 'instructor' role for the teacher (e.g. Department of Education and Training (ACT), 1997). The hands-on, concrete science approach matches the cognitive level of the majority of the students even into secondary school, and the practical work involved is usually fun and engaging. However, an increasing number of texts and papers dealing with matters of methodology, particularly those with a focus on meeting the needs of students with difficulties or disabilities, now acknowledge that there is also a definite place for direct and explicit teaching involving demonstration, explanation, modelling, practice and frequent review (e.g. Bell, 2002; McCleery & Tindal, 1999; Polloway, Patton & Serna, 2005). Teachers should not feel guilty if at times they do instruct students directly and if they impose a structure on the work to be done. Ritchie (2001), in addressing the needs of intellectually disabled students, suggests that there is definite value in science activities that are tightly structured by the teacher to ensure a reasonable chance of success and to avoid frustration. Mastropieri and Scruggs (1993) have wisely observed that, for some students with learning difficulties, open-ended discovery methods often place inappropriately high expectations on students' prior knowledge, metacognition and problem-solving abilities. These students can quickly become confused and learn little or nothing from open-ended activities that they do not fully understand. Some will also develop a very negative attitude toward the subject.

The following points suggest some practical and feasible ways of implementing effective science lessons for mixed-ability classes:

- At the beginning of a lesson, always spend enough time to settle students down and get them attentive. This is particularly important when moving from classroom base to science laboratory with potential distractions.

- Help students develop a positive mind-set by giving the lesson a clear aim and purpose.

- Try to assess students' prior knowledge on the topic by questioning and discussion as the first step.

- Outline briefly the activities to be attempted during the lesson, and suggest how the available time will be used. Summarise this information on the whiteboard.

- When possible, have a step-by-step pictorial summary of the main steps to be followed when doing an experiment. Refer the students to this self-help chart at regular intervals.

- Don't pass out materials until it is time for students to use them and until students have been instructed in their correct and safe use.

- Don't begin activities until you are certain that students understand fully what they are required to do.

- Establish a routine for the distribution and collection of materials.

- Establish rules regarding safety.

- Teach or review any necessary component skills before starting an activity or experiment (e.g. reading a thermometer, graphing data, calculating an average).

- Encourage self-regulation and self-management (e.g. checking the lesson summary on the whiteboard, being aware of time passing, asking for help when needed, clearing away materials).

- Identify 'main ideas' as the lesson progresses and write these on whiteboard.

- Use peer-assistance methods when possible, particularly for practical activities.

- Keep worksheets simple and direct. Avoid extraneous detail and complex instructions.

- Use metaphor and analogy whenever possible to help explain scientific concepts and processes in terms students can understand more easily.

- During discussion related to investigations and experiments, use clues, questions and suggestions to help support (scaffold) students' thinking.

Science in special schools

Because an understanding of basic scientific principles is important for all students, strong arguments are presented in the professional literature and in departmental policies to support the teaching of science (albeit in a modified form) within the curriculum of special schools for students with intellectual disability (DfES 2002b; Fenton, 2002; Marvin & Stokoe, 2003; Polloway, Patton & Serna, 2005). Science in special schools has great potential appeal for these students due to the many enjoyable 'hands-on' and multisensory learning opportunities that the subject can provide. A practical approach to science has also been found much more effective than a textbook approach for students with disabilities and behavioural difficulties (McCarthy, 2005).

Simple science activities can encourage low-functioning students to participate and communicate; such activities are particularly useful for engaging students' attention and curiosity, stimulating their powers of observation, extending their concentration span, developing logical reasoning and helping them make better sense of the natural and technological world (Fisher, 2002; Ritchie, 2001). Even more importantly, science activities can enhance students' further cognitive development (Henderson & Wellington, 1998).

Much of the information provided in this chapter applies equally to students with learning difficulties in mainstream schools and to those in special settings. Many of the teaching suggestions can be adapted for work with intellectually disabled students, provided even greater emphasis is placed upon carefully structured activities with relevant objectives. The approach to be used in special schools must aim to stimulate students' cognitive and sensory development and encourage verbal and non-verbal communication.

Anticipating areas of difficulty

It is helpful if teachers of science, whether in mainstream or special school, are aware of the main obstacles encountered by students with learning difficulties. For example, Bancroft (2002) identifies communication problems, weaknesses in literacy and numeracy, poor reasoning ability and deficiencies in memory as common problems. The following sections explore and explain some of these areas of difficulty. Although it is impossible to eliminate all problems, particularly when operating a science lesson with a mixed-ability class, teachers can to some extent minimise the impact of learning difficulties by careful planning before the lesson and flexible organisation and management during the lesson.

Attention and concentration problems

As indicated in Chapter 4, the first requirement for learning anything new is *careful attention to task*. Science teachers should place great importance on establishing attending behaviour before commencing a lesson, and then do all they can to maximise and maintain students' attention during the lesson. Unless students' attention is held, very little learning will occur. There are many potential distractions for students in science laboratories, and students may not be looking or listening carefully when instructions are given or demonstrations take place. Practical suggestions for enhancing attending behaviour are listed in Chapter 4.

Memory and recall problems

It is perhaps the most characteristic feature of students with learning difficulties that they don't remember information that has been covered in lessons. This situation can frustrate the teacher: 'But they knew it yesterday! How can they forget it so quickly?' The most common reasons why students can't recall information include:

- They were not attending sufficiently at the time the information was presented (e.g. daydreaming, preoccupied or anxious, lesson exceeded concentration span).

- They were not interested because the information was of no personal significance.

- The new information did not connect with anything the students already knew, so it did not become assimilated and accommodated firmly into an existing mental schema.
- The information was not understood, or understood only superficially, so it did not enter semantic memory.
- The information was not reviewed, practised or applied sufficiently.
- The students did not have effective strategies for helping them store and recall information (e.g. they did not use mnemonics or other 'tricks' to aid recall).

To help students retain and recall information, the following suggestions may be useful for the teacher of science:

- Help students feel emotionally at ease and comfortable in your science lesson.
- Link new concepts and information closely and realistically with what students already understand in their daily lives.
- Review and rehearse important information and skills frequently.
- Highlight relevant information that you expect all students to remember.
- Help students develop strategies (e.g. mnemonics) for storing and recalling information.
- Use graphics, advance organisers and key words displayed visually to reinforce and supplement information presented orally.
- Encourage students to design their own memorable diagrams or other pictorial recordings as *aides-mémoires*.

For more detailed coverage of the relationship between learning and memory, see Westwood (2004) or Bristow, Cowley and Daines (1999).

Stages of cognitive development

The teacher of science needs to be aware that one cannot force conceptual development to occur in advance of a child's stage of intellectual development, and therefore content of the curriculum must be geared to students' current cognitive level. Learning difficulties arise if concepts, ideas and relationships presented in science are beyond students' reasoning ability and experience. Activities in science for young children and for students with learning difficulties must be developmentally appropriate and reality-based if they are to be understood.

The theories of Piaget (e.g. 1963; 1971) provide a helpful point of reference for teachers seeking an understanding of cognitive development. Piaget's theory is an attempt to describe the ways in which children's perceptions and thinking change qualitatively as they get older and more experienced. The types of scientific knowledge that can be processed and understood fully by students are largely determined by the stage of cognitive development they have reached.

According to Piaget, there are four recognisable stages in cognitive development, namely: (1) the sensorimotor stage; (2) the pre-operational stage; (3) the concrete-operational stage; and (4) the formal operational stage. The sensorimotor stage usually occupies the years from birth to age two+ years, but may extend far beyond this age for children with intellectual disability. The preoperational stage commonly extends to age seven years, but well above seven years for students with significant learning difficulties. The concrete operational stage was first thought to extend from seven years to age 12, but it is now believed that it may extend to at least 16 years or above in curriculum areas such as mathematics and science (Collis & Romberg, 1992; Lovell, 1978). The final stage of development is termed the formal operational stage, beginning in adolescence and extending into adult life. At this stage abstract reasoning becomes feasible for the majority of learners (Ormrod, 2000).

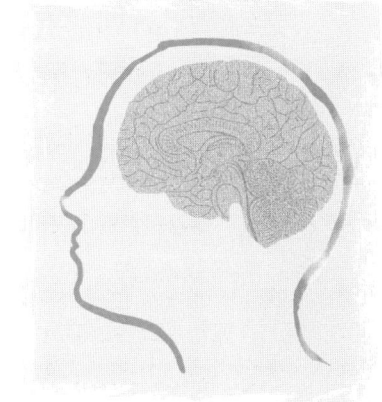

In special schools for students with developmental delay, teachers may be working with students who have severe degrees of intellectual disability and additional handicaps. The cognitive abilities of some of these students are still at the *sensorimotor stage*. Their cognition is related almost entirely to sensory perception and motor activity. Through their own actions, reactions and explorations they begin to understand object permanence and simple causality (e.g. 'If I push the plate it will fall from the table'). They also develop awareness of the regular features of the physical environment around them (shapes, textures, sounds, relative size of objects, temperature variations and weather changes) and begin to develop spatial awareness. For students with severe and complex disabilities at the sensorimotor stage of cognitive development, simple science experiences must be vivid and presented through multisensory methods (Hemmens, 1999; Marvin & Stokoe, 2003).

Schmedding (2001) reminds us that many students in primary school, particularly young children below age seven and some older children of lower ability are still at the *pre-operational stage* of cognitive development. Children at this stage have great difficulty carrying out logical thinking, even when related to phenomena they can directly observe. They are unable to make deductive inferences, using instead naïve intuition when interpreting cause and effect relationships (McInerney & McInerney, 2002). In science lessons, children at this stage of development may have problems observing salient features within a simple experiment and drawing from them a reasonable conclusion, even when directed and cued by the teacher.

Some of the typical misconceptions evident later in students' scientific knowledge often begin at the pre-operational level because children sometimes misinterpret information and attribute events to incorrect causes. An example is the case of a preschool child who thought that when the parent came to the bedroom each night and switched off the light they were actually *switching on the darkness*. Children's misconceptions have usually been constructed from their everyday experiences, and represent an attempt to explain how the world works in simplistic terms. Teachers who believe that misconceptions can be

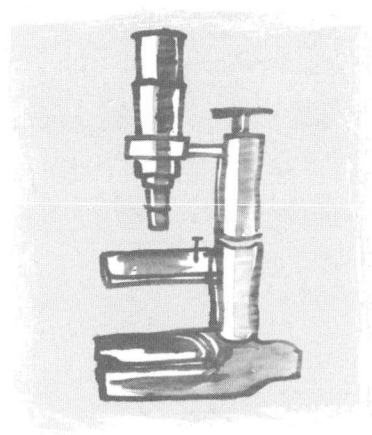

removed simply by explaining and pointing out children's mistakes will meet with little success. Most misconceptions are usually resistant to change and may inhibit further conceptual development (Rowell, Dawson & Lyndon, 1990). There is a wonderful article by Watson and Kopnicek (1990) describing the misconceptions of some nine-year-old children in relation to 'heat'. Their parents have said things to them such as, 'Put on your *warm* coat', so the children have come to believe that a coat has the property of *generating* heat. The article describes the way in which their teacher encouraged them to carry out simple experiments in the classroom to discover if clothing does indeed generate heat. Even after many instances showing that clothes remained at room temperature, the children still found it almost impossible to abandon their original erroneous belief; they still wanted to believe that objects like clothes and blankets generate heat.

Misconceptions in science are not confined to children at the pre-operational stage of development but can occur in students of any age and ability (Committee on Undergraduate Science Education, 1997). Common misconceptions include, for example, believing that clouds come from somewhere above the sky; that air is weightless; that heavy objects fall faster than light objects; that magnets can pick up any metal object; that the planets, sun and moon revolve around the earth; that the earth is the largest object in the solar system; or that a human sperm is the size of a tadpole (because that is the impression left by the diagram in a textbook).

If we move beyond the pre-operational stage, we find that even more students in primary classes are at the concrete-operational stage of cognitive development. At this next stage, although the students are slightly more capable of logical thinking, they really only understand easily the things they can see, manipulate and experience first-hand. This is why practical, hands-on activities in science are so valuable for them, and why purely chalk-and-talk and textbook science lessons are almost bound to fail. If practical activities are age-appropriate, carefully structured and presented in clear steps, children at the concrete-operational level can engage in them successfully and develop interest, confidence and belief in their own self-efficacy (DfES, 2002b). It is said that abstractions of all kinds can gradually make their appearance as children mature and develop an ability to handle explanations that are more complex and abstract (AAAS, 1993).

For students at the pre-operational and concrete-operational stages it is essential to try to link new ideas in science with the children's everyday experiences. Good and Brophy (2003) remind teachers that the subject is made more meaningful and accessible by using real-life examples and everyday applications of scientific principles. Similarly Polloway, Patton and Serna (2005) indicate that science topics should always be made relevant to students' lives, and connections should be made between concepts encountered in science and other subject areas in the curriculum. They also suggest that teachers spend more time on fewer topics, because if the pace at which new concepts are introduced outstrips students' rate of learning, they either turn away from the subject in frustration or simply engage in superficial learning (Hackling, Goodrum & Rennie, 2001; Hanrahan, Cooper & Russell, 1997).

It is pertinent to point out that many students in secondary schools, even in the upper grades, are still operating at the concrete level in science (Lovell, 1978). This is one of the main reasons why they often find science a difficult subject to understand, and why they may resort to approaches such as rote memorisation of definitions, exemplars and laws, rather than persevering with attempts at deeper learning.

In any mixed-ability class there will be students at various stages of cognitive development and with varying degrees of competence in language, literacy and numeracy. For this reason it is often necessary for teachers to provide a variety of activities and resources that are differentiated to some extent to match students' capabilities (DfES, 2002a).

Teachers may find the following points useful in relation to planning science lessons compatible with students' cognitive level:

- For students with complex learning difficulties, always use multisensory methods to present exploratory science activities that are stimulating, vivid and enjoyable.

- Try often to use spontaneous activities and interests of the students as starting points for exploration.

- Provide clear links between principles and concepts covered in science lessons and real-life situations or applications familiar to the students.

- Use effective questioning strategies that help students focus on relevant data and stimulate thought.

- Encourage students to describe and discuss their thoughts and observations.

- Structure practical learning activities carefully so that students are clear about their purpose and know exactly what they are required to do.

- Allow students to employ developmentally appropriate recording systems (e.g. pictorial) to minimise problems with symbolic representation and abstract notation.

- Identify and consolidate 'main ideas' as the lesson progresses, and write these on the whiteboard.
- Check carefully for misconceptions concerning what a particular activity has revealed.

Language difficulties

Language is central to thinking, communicating and problem-solving in all subjects because language helps a learner reflect upon and describe observed relationships, ask salient questions of peers or teachers, and understand instructions and explanations given by others. Henderson and Wellington (1998) state that the greatest obstacle to enjoying and learning science for many students is the language barrier. Unfortunately, it is fairly characteristic of students with learning difficulties that they are not particularly proficient in expressive or receptive modes of language (Bell, 2002). When students disengage from a task it is often because they no longer understand what the teacher is explaining or asking.

It is easy to understand that unfamiliar and complex technical terms present an obstacle to communicating in science (Bahar, Johnstone & Hansell, 1999), but even some common non-technical words used frequently in questioning or instructing are not understood by some students with learning difficulties (e.g. *assemble, tighten, reduce, compare, interpret, devise, adapt, modify, record*). In addition, Churton, Cranston-Gingras and Blair (1998) suggest that some students have great difficulty remembering new vocabulary associated with various experiences or observations. Words like *energy, force, pressure, velocity, power, mass, weight* and *density* appear to give particular difficulty to many students.

To enhance language development in science, frequent use must be made of group and whole-class discussions. Through describing, talking, listening and answering questions, students clarify their own thinking and help other students achieve better understanding (Henderson & Wellington, 1998). Discussion, involving the sharing of ideas among students and with the teacher, is essential to help foster deeper learning and language enrichment.

The importance of questioning to stimulate students' thinking was stressed in Chapter 4. Effective questions in science are those that help students reflect upon their observations, clarify their thinking and reveal to the teacher the depth of their understanding (DfES, 2004). The teacher may need to help students become more explicit in their responses by saying, for example, 'I'm not sure I understand . . . tell me more' or 'OK. How might you explain that to a younger student?' Bell (1999) also reminds teachers of the need to allow adequate wait time for thinking, after asking a question.

The following points may help science teachers address and reduce problems related to language difficulties:

- Anticipate the vocabulary and comprehension problems some students will have in relation to a new topic.

- Teach the language of the subject as well as the subject matter.

- Preteach and overlearn essential terms associated with particular topics, themes or activities. Ideally, present the meaning of the word, the pronunciation of the word and the recognition of the word in print.

- Use language at a level students can readily understand when asking for their ideas, giving instructions, providing explanations, asking and answering questions and summarising points.

- Simplify complex directions by breaking them down into several steps.

- Ask students to repeat instructions they have been given.

- Avoid issuing multiple instructions.

- Routinely repeat instructions in a clear and positive manner.

- Differentiate your questioning according to your knowledge of students' ability.

- Allow adequate wait time for students to respond to your questions.

- Constantly check for understanding.

- Encourage students to ask questions and give opinions.

- Encourage students to tell you when they do not understand something.

- Recognise the potential value of science activities to stimulate students' language skills.

Literacy problems

Another common characteristic of students with learning difficulties is that they have problems with reading, writing and spelling. In reading they may have difficulty at the basic level of word recognition and decoding, or they may have problems in understanding what they have read, even if they can identify the words on the page. These problems were discussed in Chapter 5. In science, this weakness manifests itself most frequently as an inability to read and understand textbooks, blackboard notes or instruction sheets (Cawley & Parmar, 2001; Churton, Cranston-Gingras & Blair, 1998). The readability level of most science textbooks, particularly in secondary schools, is not usually geared to the reading level of weaker readers. The more dependent the learning of science is on written assignments, textbooks and worksheets to convey information, the more likely it is that students with literacy problems will find the subject difficult and frustrating (Hanrahan, Cooper & Russell, 1997).

Teachers of science can approach the literacy challenge in three main ways:

- They can reduce the emphasis placed on reading as the primary method for gaining information in favour of a more oral approach, using participation and discussion (Cawley & Parmar, 2001). Scruggs et al. (1993) report that students much prefer practical activities over textbook or chalk-and-talk lessons. Students with learning difficulties tend to remember more and maintain better attention in such lessons.

- Where textbooks or other print material must be used, the teacher might prepare modified versions of the text and provide differentiated instruction sheets with simpler vocabulary and shorter, less complex sentences (Lovitt & Horton, 1994). Ritchie (2001) and Versey (1993) advocate differentiation in science, with tasks set and questions posed to match students' different ability levels as closely as possible. The aim is to reduce failure rate and build students' confidence by giving tasks and materials they can manage independently. However, there is some evidence that adolescent students in particular do not like being given materials that appear much easier than those given to other class members (Hall, 1997; Lo, Morris & Che, 2000). Bancroft (2002) is adamant that materials given to students with learning difficulties should not lower their self-esteem and should be as close as possible in appearance to the mainstream materials used in the class. Differentiation, although advocated in most curriculum guidelines, is clearly inappropriate if it highlights a student's difficulties in a very public manner.

- The old truism that 'all teachers are teachers of literacy' is indeed true. Teachers of science can recognise the valuable role they can adopt in using opportunities in science lessons to increase students' reading, writing and spelling skills (Cawley & Parmar, 2001). For example, if a textbook is used they can adopt a 'guided reading' approach (see Chapter 5) and can help students interact more with the text to deepen their understanding. They can preteach vocabulary and ask questions that will support comprehension of the text. Henderson and Wellington (1998) recommend building a science vocabulary word-bank, adding to it and reviewing it over time. All teachers of science should compile a list of essential vocabulary associated with the subject, and should ensure that all students understand these words, recognise them in print and write them correctly.

To help students with severe writing difficulties, the teacher of science can reduce the demands of writing tasks by:

- Preparing whiteboard or overhead projector summaries using cloze procedure, where key words are omitted and must be supplied by the student.
- Providing sentence beginnings so the student can complete the sentence (e.g. 'When two bar magnets are brought together, the unlike poles will _____.').
- Using true/false items in a summary sheet.
- Using multiple-choice questions.
- Using short-answer items.

Other practical ideas for adapting students' notes and worksheets can be found in the book *Modifying Schoolwork* by Janney and Snell (2004).

To reduce the problems associated with literacy difficulties, teachers may:

- Look for (or produce) alternative text materials at a simpler readability level for some students.
- Actively help students read and interpret the textbook.
- Spend adequate time ensuring that all students can recognise key terms in the textbook, on the blackboard or on the computer screen.
- Help students learn correct spelling of essential words in science.
- Read instructions aloud for the benefit of students with reading difficulties.
- Use clear simple English in all worksheets and instruction cards.
- Try to reduce the overall use of print materials when other methods of presenting information are available.
- Recognise that helping students read and write in the science genre can greatly enhance the general development of their literacy skills.
- Ensure that students with learning difficulties are not handicapped by assessment methods that rely too much on good reading and writing ability.

Additional accommodations for secondary school students with learning difficulties include appointing a scribe or note-taker for the student, allowing students to borrow notes of other students, providing lesson summary notes as handouts, audio recording of class sessions, and providing extra time for written work and tests.

Numeracy problems

Gross (2005, p.13) points out that, 'Very little modern science can be done without calculation' — so difficulty with basic arithmetic presents a problem for any student learning science (Bahar, Johnstone & Hansell, 1999). Even the simplest of scientific explorations usually require students to count, measure, calculate, compare and record numerical data. Unfortunately, an area of weakness for many students with learning difficulties is numeracy (Micallef & Prior, 2004). These students often have poor intuitive 'number sense' and lack a natural feeling for quantitative relationships. They seem unable to recognise when a result they obtain is not feasible, and may have difficulties selecting the correct arithmetic process to use when, for example, finding numerical differences or calculating averages. Their ability to carry out basic arithmetic and solve algorithms accurately is often poor, and they lack confidence in their own ability to obtain correct results.

Problems with basic mathematical skills may be due to insufficient teaching and inadequate practice in the earlier years. The difficulties may also be related to weakness in generalising number skills from the mathematics class to new contexts and applications. Difficulties in generalising and transferring knowledge and skills are well-recognised in students with learning difficulties (Churton, Cranston-Gingras & Blair, 1998; Ritchie, 2001). However, part of the difficulty for students just entering the concrete-operational level may arise because they are not confident with certain aspects of symbolic representation. Signs such as $\pm$, Σ, $>$ and $<$ seem difficult for them to understand, and they may also have problems grasping ideas of ratio and scale — they may, for example, have difficulty selecting an appropriate scale when graphing results.

Some hints for a teacher wishing to minimise the impact of students' difficulties in numeracy include the following:

- Do not make too many assumptions concerning the students' basic mathematical skills.

- Explicitly and clearly demonstrate appropriate ways of recording and analysing quantitative data.

- Provide worked examples for any calculations that need to be done on the blackboard or on a wall chart.

- Provide worked examples of effective recording methods.

- Encourage students to present results in a format that is meaningful to them, such as pictorial recording.

- Provide learning aids such as calculators or number charts to reduce the load involved in calculating.

- Teach students easy methods for checking the feasibility of a numerical result.

- Encourage peer assistance during the activities and analyses.

Social and behavioural problems

Ritchie (2001) suggests that group work in science can be of benefit in helping students develop interpersonal skills and communication. Some students with learning difficulties have problems working easily as a member of a group and may be distracted by irrelevant happenings. They may argue with peers or disrupt the activity. Teachers need to give careful thought to the composition of working groups and avoid putting together any students who may be antagonistic and incompatible. Sometimes it helps to give a student with difficulties a particular duty or role within the group, then make clear to him or her exactly what is required. Even when students are not troublesome in the group there may still be a tendency for them to waste time and sit on the sideline, leaving the work to others.

Clumsiness

A few students with learning difficulties have genuine coordination problems and are seen to be clumsy when carrying out fine motor actions (Mastropieri & Scruggs, 1993). In the science lesson this can manifest itself as a tendency to drop and spill things frequently, or to have difficulty in assembling apparatus. The science teacher needs to recognise this difficulty and adopt a tolerant attitude toward minor clumsiness. Allocating a partner and helper can often reduce this problem.

Weak self-efficacy

Due to a history of failure many students have developed a poor view of their own efficacy in any learning situation. They expect to fail when confronted with any new task. In science they may be reluctant to attempt certain activities or may leave the work to other members of the group. The science teacher needs therefore to be encouraging and supportive in order to strengthen the student's willingness to have a go. If tasks are structured and students gain success, science becomes an area where their positive self-efficacy beliefs can be increased.

It is becoming increasingly common to find that there is additional adult help available during science lessons (DfES, 2004). In most secondary schools, for example, there is a laboratory assistant or technician, and at primary level there may be a member of staff with the role of learning support assistant or classroom aide. Assistants in class can help students understand and participate in the tasks they are set, interpret instructions, and communicate and record their ideas.

Poor self-management

Some students with learning difficulties are poor at planning, regulating their use of time and preparing themselves for activities. Sometimes pairing these students with a suitable partner can help in the short term to ensure the smooth flow of the lesson; the long-term aim, however, will be to help these students become better organised. For some students it is important that the teacher begins each practical lesson by outlining on the whiteboard the activities to be undertaken, the materials required and the approximate time allocation for each activity. Students can refer to this summary as the lesson progresses. Mastropieri and Scruggs (1993) refer to 'self-monitoring sheets' on which students can check off each step in an activity as it is completed.

Poor-quality instruction

Although the ideal approach to science teaching as described in official documents and guidelines involves practical investigation, collaborative learning and relevant curriculum content, the reality is often very different (Hackling, Goodrum & Rennie, 2001; Weiss et al., 2003). Too often teachers employ routine chalk-and-talk lessons with little practical activity involving the students. The curriculum content may be covered too rapidly and too theoretically, and students end up disliking the subject. This may be particularly the case when a non-specialist teacher is required to teach science but has no deep subject knowledge and even less pedagogical content knowledge related to science.

On the other hand, recent studies have shown that an appropriate degree of high-quality direct instruction — for example, explaining key concepts and demonstrating how to set up particular experiments — improves the learning outcomes in many science classes and particularly for students with poor learning skills (Cavanagh, 2004; Tweed, 2004). A study of classroom teaching of mathematics and science in the US found that effective lessons did not always involve discovery learning but rather used topics and teaching strategies (particularly effective questioning) that engaged and motivated the learners (Weiss et al., 2003). As we have seen already in relation to the teaching of reading and basic mathematics, an approach that combines teacher-directed methods with appropriate student-centred activity probably produces the best results also in science.

Useful resources

Science education (students with learning difficulties)

DfES (Department for Education and Skills) (2004). *Science module: Induction training for teaching assistants in secondary schools.* London: DfES. Viewed 3 February 2006, (http://www.teachernet.gov.uk/_doc/7133/Sci_Sec.pdf). Although designed mainly for the use of teaching assistants, teachers will also find much of value in this material.

ERIC EC (ERIC Clearinghouse on Disabilities and Gifted Education) (2003). *Teaching science to students with disabilities.* USA: Hoagies' Gifted Education Page. Viewed 3 February 2006, (http://www.hoagiesgifted.org/eric/faq/science.html)
This resource, together with those listed below, will provide teachers with many practical suggestions for teaching science to students with intellectual and other disabilities. The main pedagogical principles apply, regardless of whether science is being taught in the context of a special school or in a mainstream inclusive classroom.

Inclusive Science and Special Education Needs (n.d.). Hatfield, UK: Association for Science Education. Viewed 3 February 2006, (http://www.issen.org.uk).

Inclusive science and special educational needs resources (2003). CD-ROM. Hatfield, UK: Association for Science Education.

NCC (National Curriculum Council) (1992). *Teaching science to pupils with special educational needs.* York: NCC.

QCA (Qualifications and Curriculum Authority) (2001). *Planning, teaching and assessing the curriculum for pupils with learning difficulties: Science.* London: QCA. Viewed 3 February 2006, (http://www.nc.uk.net/ld/Sc_content.html).

Rakes, T.A., Choate, J. & Stringer, G.L. (2004). Essential science: Relevant topics, process, and strategies. In J. Choate (ed.) *Successful inclusive teaching: Proven ways to detect and correct special needs* (4th edn, pp.312–343). Boston: Allyn & Bacon.

SESD (Science Education for Students with Disabilities) (2006). New York: SESD. Viewed 3 February 2006, (http://www.sesd.info/index.htm)

8 Teaching social studies, history, geography and environmental education

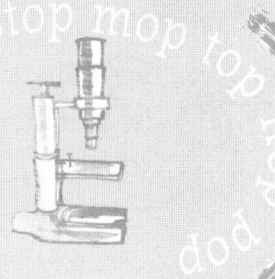

> Social studies [and related subjects] promote informational skills and values development that contribute substantially to an understanding of human diversity, societal complexity, and general world knowledge (Polloway & Patton, 1997, p.385).

Social studies, history, geography and environmental education are presented together in this chapter because effective teaching practices tend to be very similar across all four subjects, and potential problems in learning are basically the same in the four areas. Presenting these subjects together is also justified because although history and geography are still identified as separate subjects within the National Curriculum in Britain, many other education systems adopt a more integrated approach, at least until the secondary school years. In the US, for example, social studies as a school subject integrates not only geography and history but includes coverage of civics and government, ethics, social justice, economics, philosophy, psychology, religion and sociology.

Typical social studies themes for which teaching standards have been specified in the US include: culture; time, continuity and change; people, places and environment; individual development and identity; individuals, groups and institutions; power, authority and governance; production, distribution and consumption; science, technology and society; global connections; and civic ideals and practices (National Council for the Social Studies, 1994). The guidelines created by the New York State Education Department (NYSED) are fairly typical of other curriculum guidelines in suggesting that:

> Social studies classes help students understand their roots, see their connections to the past, comprehend their context, recognise the commonality of people across time, appreciate the delicate balance of rights and responsibilities in an open society, and develop the habits of thoughtful analysis and reflective thinking (NYSED, 2005, p.2).

In Australia, social studies, history, geography and environmental education are subsumed under the subject Studies of Society and Environment (SOSE) (Curriculum Corporation (Australia), 1994b). This domain of study is organised into five conceptual areas: time, continuity and change; place and space; culture; resources; and natural and social systems. Emphasis is placed on using active inquiry methods that ensure maximum student interest and participation. Typical guidelines in Australia state that:

> Studies of Society and Environment helps young people understand how people's life experiences are the result of particular social, cultural, economic and environmental relationships that characterise communities at particular times and places. The values, concepts and skills of the learning area are drawn from a range of traditions of inquiry (Queensland Department of Education and the Arts, 1998, p.1).

The topics and themes selected as content material for social studies and related subjects such as environmental education should stimulate students' investigative and problem-solving skills, such as observation, interpretation, clarification, comparison, prediction, inference, critical thinking, judgment, decision-making and the generation of new ideas (Pressley & McCormick, 1995; Smith & Smith, 1997). Social studies teachers draw upon many different disciplines in order to teach students the content knowledge, intellectual skills, attitudes and values necessary for fulfilling the duties and responsibilities of citizenship (National Council for the Social Studies, 2005).

From the information above, it can be seen that social studies comprises an important part of the curriculum for all students. Inclusive education principles suggest that all students, regardless of degree of ability, disability or learning difficulty have a right to participate in social studies, history, geography and environmental education.

Social studies and related subjects: aims and purposes

The primary purpose of teaching social studies is to help individuals gain relevant knowledge and attitudes to enable them to make informed decisions as members of society (National Council for the Social Studies, 1994). Experiences provided within this area of the curriculum should challenge students to respond creatively, both as individuals and with others, to important local and world issues.

In carrying out learning activities typically provided in social studies and similar courses, students are also practising and applying basic skills in

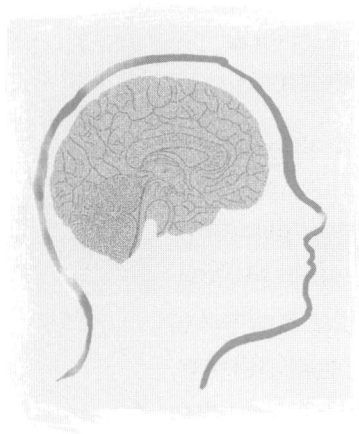

communication, language, literacy, numeracy and ICT, as well as gaining new information, developing new skills and acquiring beliefs and attitudes related to particular issues. Students are also exercising important study skills such as critical thinking, searching multimedia sources for information, note-taking, data-recording, interviewing, summarising, composing, editing, illustrating and communicating ideas (Janney & Snell, 2004; Mackintosh, 2005). Teaching in social studies aims to promote 'thoughtfulness' in students, meaning that the approach to topics studied should arouse students' curiosity and enhance their ability to reflect critically on ideas; check information; weigh evidence; and consider alternative solutions, plans or possibilities (Good & Brophy, 2003; Leming, 2003; Newman, 1991).

History

Where history is taught as a separate subject, the aim is to help students develop an interest in, and an understanding of, the past and how the past is reflected in aspects of the world of today. According to von Heyking (2004, p.2), '[History] is a form of inquiry that helps us construct an understanding of our own lives (individually and collectively) in time'.

For children in primary and special schools, learning history should be much more than the memorising of interesting facts and stories about famous people and events. In England, the guidelines issued by the Qualifications and Curriculum Authority (QCA, 2001e) suggest that when studying history, students with learning difficulties will learn about the recent past, the more distant past, and how their own role in family and community has changed. According to the QCA, studying history offers students with special educational needs opportunities to:

- Develop knowledge and understanding of the sequences, routines and chronological patterns that make up their world.

- Develop an understanding of their personal history alongside understanding about events in the world and what shapes them.

- Develop knowledge and understanding of how people lived in other times and how those times were different from today.

- Experience a range of representations of the past.

- Use a range of evidence to find out about the past (QCA, 2001e, p.2).

Von Heyking (2004, p.11) states:

> Understanding that the actions of people in the past have an impact on us today, and appreciating that our actions will have consequences for future generations is history teaching's essential contribution to citizenship education.

Ormrod (2000) reminds us that understanding history requires the application of several important abilities and processes, namely 'thinking historically', understanding the nature of historical time, drawing inferences from historical evidence, making connections, identifying change and cause–effect relationships among events, and recognising that historical figures are not fictional but were real people. Some of these abilities and processes are problematic for students with learning difficulties (Martin & Gummett, 2001).

Geography

Where geography is taught as a separate subject, the aim is usually to extend students' awareness and interest in their surroundings by identifying and exploring features of the local and wider environment (Gummett & Martin, 2001). According to QCA (2001d), when environmental education issues are incorporated  in geography, there are four main areas to be developed: geographical inquiry and skills, knowledge and understanding of places, knowledge and understanding of patterns and processes, and knowledge and understanding of environmental change and sustainable development. Effective learning in geography involves such abilities as identifying interrelationships among people and their environments, appreciating cultural differences and understanding that maps are symbolic and scaled representations of real places (Ormrod, 2000). Again, some of these abilities are problematic for students with learning difficulties. In addition, field trips and excursions that form an essential part of effective geography teaching can present difficulties for some students with disabilities and for their teachers (Healey et al., 2002; Heimlich & Daudi, 1996). Advice on managing the logistics of field trips for students with disabilities is provided by Chalkley and Waterfield (2001).

Environmental education

In schools where environmental education is treated as a subject in its own right, the focus is on learning *in* the environment, *about* the environment and *for* the environment. The study of this subject is not simply confined to gaining information about flora, fauna and natural resources, but stresses also the development of positive attitudes, values and behaviours associated with environmental protection, sustainable development and responsible lifestyles. Students begin to understand how the environment affects people's lives and how people in turn affect the environment. They observe ways in which the physical world is modified by human activity, largely as a consequence of the ways in which human societies value and use natural resources (NYSED, 2005).

Key concepts commonly explored in environmental education (often using a problem-based approach) include natural resources management, sustainability, interdependence, biodiversity, and personal responsibility for positive and protective action. Students are helped to clarify and articulate their own attitudes, values and beliefs concerning their place in, and responsibility for, the environment (Department of Education and Training (ACT), 1994). They are also involved in practical projects that involve, for example, energy conservation, recycling of materials, revegetation and reduction of pollution.

Environmental education is most commonly embedded within geography or social studies programs, but in some schools it is included as an important component of 'outdoor education'. Outdoor education itself has clear links also with the physical education curriculum, particularly where camping and expeditions are involved (OfSTED, 2004). Outdoor education has been a strong feature in the curriculum of good special schools for many years, and contributes enormously to the socialisation, self-help and independence training for students with disabilities.

In summary, it can be seen that social studies, history, geography and environmental education provide abundant natural opportunities for students with or without special educational needs to enrich their vocabulary, develop their literacy and numeracy skills, apply observation and inquiry skills, and extend reasoning and concept development while studying interesting, relevant and enjoyable topics in practical ways (OfSTED, 2005b). All four areas of study lend themselves to an issues-based or problem-based learning approach (King, 2001; Lee, 2001). There is also great scope for individual or collaborative project work and cooperative learning. Implementing a resource-based approach to these subjects enables students to use, for example, books, videos, CD-ROMs, real objects, models and community publications to study social or environmental issues, or use genuine artefacts and old photographs to study local history. In this area of the curriculum, use can be made of classroom learning centres, as well as the school library, resource room and ICT facilities (Okolo & Ferretti, 1996).

Approaches to teaching and learning

Polloway and Patton (1997) provided a useful critique of approaches to teaching social studies and related subjects, with particular reference to students with learning difficulties. They contrast the 'traditional approach' (based on teacher-talk, textbooks, supplementary resource materials and teaching aids) with the 'inquiry approach' and what they term a 'balanced approach'.

The traditional approach, using direct information input from the teacher and published texts, places less pressure on teachers in terms of resource preparation and presentation time. The scope and sequence of content in standard programs and textbooks has usually been designed carefully and has often benefited from creative input by curriculum experts. High-quality books and materials (e.g. supplementary videos, CD-ROMs) can be quite stimulating in terms of illustrations and animations. However, published programs and texts often present too many concepts too quickly, and the readability level of the print material is often beyond the independent reading competence of students with learning difficulties. There is also a lack of flexibility that prevents or restricts teachers' opportunities to capitalise on events of the moment.

The inquiry approach, based mainly on constructivist learning principles, is favoured in most contemporary curriculum guidelines. For example, information from Regina Public Schools and Saskatchewan Learning (2003, p.1) suggests that, using an inquiry approach, students 'learn how to learn':

Skills such as careful observation, reasoning, critical thinking, and the ability to justify or refute existing knowledge are developed. Creative thinking and curiosity is stimulated. Student motivation and self-esteem increases with the control over their learning. Learning becomes fun.

The approach is potentially much more flexible and student-centred than the traditional method, and can easily link content to students' genuine interests, draw on their knowledge of current and local events in the news and help develop the study skills described previously. Good and Brophy (1997) conclude that an investigative approach to social studies and related topics can encourage higher order thinking and problem-solving across all ability levels. King (2001, p.4), when advocating a problem-based learning approach for environmental education, observes that, 'Learning through problem solving is much more effective for creating in a student's mind a body of knowledge that is usable in the future, than didactic traditional methods of teaching'.

In recent years in Britain, Australia and the US, the emphasis has been on adopting the inquiry approach to involve students directly in handling and presenting authentic data, participating in discussions, role-playing, problem-solving and decision-making (Burkill et al., 1999; Smith & Smith, 1997). Engaging in such activities broadens students' range of independent and collaborative learning strategies, as well as creating natural opportunities for illustrating, calculating, writing and note-taking. However, the inquiry approach does place much more pressure on teachers to locate and gather necessary materials and to devise and manage appropriate classroom activities. There is also a danger that the program becomes a series of interesting but largely unrelated topics, because it is very difficult to establish a logical sequence in the content covered week by week when themes and topics are chosen simply in response to children's interests of the moment, or to address current issues and events.

From the viewpoint of students with learning difficulties, a definite disadvantage in the inquiry approach is that it requires learners to have sufficient initiative and effective study skills to work independently at times. The types of critical and reflective discussions that ideally occur in effective inquiry approaches are actually very hard mental work for students and also very hard organisational work for teachers (Pressley & McCormick, 1995). As Polloway and Patton (1997, p.390) remark:

> A totally inquiry-oriented approach for teaching social studies to learners with special needs must be used with caution [and] a certain amount of structured instruction that helps students make connections across topics and concepts is probably necessary.

Finally, it is noted that not all students like to learn through an inquiry or problem-based approach (although most do), and not all students like collaborative group work (Lee, 2001).

The balanced approach could more accurately be termed the 'combined' approach. It uses many of the books, resources and topics typical of a traditional approach but introduces more investigative activities, field work and projects centred on current or local issues and themes. At appropriate times, teachers can apply basic principles of explicit instruction — such as reviewing previous learning, specifying lesson objectives and expectations, discussing the importance of the topic, providing advance organisers and preteaching new terminology. Students with learning difficulties appear to benefit greatly from such measures (Hudson, 1996).

Poor-quality teaching

According to the National Council for the Social Studies (2005), social studies teaching is most powerful when it is meaningful, integrative, value-based, challenging and active. The program content should relate to the age, maturity and interests of students and should help them connect social studies concepts and principles to their own lives. Good and Brophy (1997) describe less effective teachers as being too ready to present ideas ready-made to the students, rather than encouraging students to contribute what they know and think. Less effective teachers tend to talk at the students rather than with students, and do not engage in much group discussion. More effective teachers use questions that elicit original thinking, and ask students to respond to one another's ideas. Effective teachers often model the characteristics of a 'thoughtful' person — for example, checking information, weighing evidence or considering alternative options. When students seem confused, effective teachers provide additional explanations or demonstrations, and they stay with a student until material is understood.

Students are likely to experience difficulties in learning social studies and related subjects if the quality of teaching is poor or if resource materials are scarce or inappropriate. In relation to quality of teaching in general, Darling-Hammond (2000) observes that the effect of poor teaching on student outcomes in any subject area is both debilitating and cumulative.

It is unfortunate that not all teachers feel confident or competent to teach social studies, history or geography, and perhaps have had the subject added to their timetable as an 'extra'. If it is not a teacher's major area of expertise he or she is likely to lack subject knowledge and pedagogical knowledge, leading to the adoption of a chalk-and-talk or textbook approach, rather than motivating students with problem-based and exploratory activities. It is pertinent to note that inspections of primary schools in the UK in 2004 led to the conclusion that geography, and the new subject 'citizenship education', were not being taught very effectively in many schools (OfSTED, 2005a; 2005b). Some teachers appeared to rely too heavily on commercially produced worksheets with mundane tasks that do very little to enrich and extend students' learning. Similar weaknesses have been observed in some American schools, where the worksheet approach amounts to little more than 'busy work' (Good & Brophy, 2003). The majority of schools observed in the UK did not ensure that the recommended amount of lesson time was made available for these important subjects, and often content within the program did not represent a sequential or progressive development of knowledge, skills and understandings. Taylor (2004) reports that students are being put off by poor teaching that focuses on memorisation of facts rather than developing an understanding of big issues.

Teaching can be made more effective in this area of the curriculum if the teacher:

- Gains the attention of all students before commencing the lesson.
- Uses graphic organisers to help students conceptualise the purpose, scope and sequence of activities around the central theme.
- Relates new information or concepts to what the students already know.
- Provides examples of new concepts that are within the students' life experiences.
- In history teaching, makes frequent use of a timeline to relate past events to today.
- Uses a rich variety of resources and activities to cater for different ability levels and learning styles or preferences.
- Provides a rubric to guide the inquiry or problem-solving process.
- Specifies the nature of the potential end products expected from an activity.
- Sets tasks and problems that are sufficiently interesting and challenging, but not so difficult that they cause failure and frustration; challenging tasks just within a student's grasp promote maximum cognitive growth (Ormrod, 2000).
- Balances the time allocation effectively among the delivery options of direct teaching, group work, discussion and individual assignments.
- Uses whole-class and small group discussions efficiently.
- Pauses at regular intervals during lessons to synthesise ideas and reach tentative conclusions.
- Varies assessment methods to allow students many ways to demonstrate what they know and are able to do.

Using group discussion effectively

Discussion is a very important learning and teaching strategy in this area of the curriculum. Students should have an opportunity to express their views, raise issues and listen to the ideas and opinions of others. Teachers should therefore try to:

- Ask questions that call for deeper thought and reflection.
- Model 'thoughtfulness' and reflection when approaching problems.

- Reveal their thinking processes to students.
- Encourage students to ask questions and raise issues.
- Ask students to explain or share their thinking.
- Encourage students to expand upon their ideas.
- Challenge students to support their views with evidence.
- Ensure that students listen with understanding to each other's explanations.
- Ensure that every student contributes to discussions.

The National Council for the Social Studies (2005) urges teachers to avoid grouping students strictly by ability within the class when investigating and discussing topics, and instead encourages the use of heterogeneous groupings to accommodate differences in students' ability through flexible teaching. Flexible teaching might include, for example, the use of peer tutoring, computer-assisted learning (CAL) and team-teaching.

The challenge of mixed-ability classes

To address the needs of students with learning difficulties or disabilities in mixed-ability classes, several authorities recommend that curriculum content, activities and materials should be differentiated according to students' abilities (Janney & Snell, 2004; Martin & Gummett, 2001). These principles were discussed in Chapter 1. Often, this simply means that graded worksheets and assignments are used, and perhaps easier reading materials are provided for some students. However, Steel and Hattersley (2005) favour an approach that is inclusive of all students rather than potentially divisive. In such an approach the students work at their own levels cooperatively while investigating the same issues or concepts and using the same materials and experiences as all other students in the class. The role of the teacher is to help students cope with material as presented, not necessarily to provide alternative and simplified tasks or modified resources. Lenz and Schumaker (1999) refer to this as 'mediated learning'. Under 'mediated learning' it is not curriculum content, objectives of the lesson or learning tasks that are changed, but rather it is the amount of direct guidance and additional support given to individuals that is varied. The amount and quality of work produced will also differ from student to student.

Differentiating lesson objectives for mixed-ability classes can be managed by planning lessons with three ability levels in mind. First the teacher writes a general objective for the essential core learning that all students will master in the lesson. Then an additional objective is written that builds on and extends the first objective and represents the next level of achievement that most students can attempt ('beyond the core'). Finally, an objective is written to identify higher level outcomes that a few students may achieve (QCA, 2001e; Westwood, 2003a).

A situation where differentiation is sometimes attempted is in relation to assignments or tasks set by the teacher. For example, when tasks are set for students with limited independent study skills, the teacher may provide for them more explicit clues and prompts within the question (for example, the exact page number on which the information can be located), and the task may be broken down into several steps. More independent learners have fewer clues and the task is more open-ended and challenging (Lenz & Shumaker, 1999).

Anticipating areas of difficulty

As described in more detail later, when an inquiry or issues-based approach is used to teach social studies and related subjects, great assumptions are made that learners will have at least adequate initiative and basic skills (language, literacy, numeracy and social competence) to engage successfully in both independent and collaborative learning situations. Any learning difficulties that arise are usually related to inadequacies in some of these areas. In addition, the 'unstructured' nature of an inquiry approach can create an additional barrier, because students with learning difficulties often do best in more structured programs, delivered by direct teaching methods (explicit instruction) (Purdie & Ellis, 2005; Mastropieri, Scruggs & Butcher, 1997; Scruggs & Mastropieri, 2003). For example, it is known that students with learning difficulties tend to do significantly better when objectives and advance organisers are clearly provided for learning, where material is broken down into smaller units of information, and where there is frequent teacher feedback and scaffolding (ERIC EC, 1999; Hudson, 1996; 1997).

Other learning problems may be related to curriculum content that attempts to introduce students too soon to complex concepts or issues that are beyond their current stage of cognitive and affective development. Some social studies programs appear to cover too many new concepts too quickly, resulting at best in superficial learning. Some educators suggest that we should reduce the amount of content to be covered in social studies and related courses, and focus in depth on fewer but highly relevant topics (e.g. Good & Brophy, 2003; Pressley & McCormick, 1995).

Some of these problems will be discussed in more detail below, but here are some points to help teachers minimise the effects of potential learning difficulties:

- Make sure that students have sufficient knowledge about a topic before you expect them to discuss it or work with it in groups, thus avoiding unproductive 'sharing of ignorance'.

- Seek out or adapt books and other print resources that are at an appropriate readability level for the weaker readers in the class.

- Use guided reading activities to help weaker readers understand the text.

- Identify and preteach new vocabulary associated with a new topic.

- Provide concrete examples of finished work that might be produced from groups.

- Monitor group work closely and give additional support or explanation when necessary.

- When questioning, allow adequate time for students to think before requiring them to respond: 'The pace of effective instruction is always such that students respond only after they have reflected' (Pressley & McCormick, 1995, p.286).

Difficulties with literacy and language

Difficulties in reading make the task of searching for information in print or from electronic sources very frustrating (ERIC EC, 1999). The frustration can lead to loss of motivation and then disengagement from the learning activity. As far as is reasonable for certain students, social studies and related subjects should be presented in ways that do not place a premium on good reading ability; where reading is absolutely necessary, the demands of the task should be reduced. Simple tactics such as enlarging the print size on overhead transparencies, using concept maps or other graphic organisers instead of lengthy written notes and preteaching the recognition and meaning of key terms can be first steps toward reducing the reading demands of the subject. At all age levels, the golden rule for teachers should be to avoid creating an information overload, particularly when providing printed handouts and when setting private reading assignments (SCoTENS, 2005).

Waterfield (2002) indicates that students with literacy problems usually need help locating and using library resources, taking appropriate notes, organising written work and completing homework assignments. They may also need extra time for carrying out assignments and tests. Similarly, for students who find writing and spelling difficult, the demands for extensive essay or report writing and note-taking should be reduced. However, note-taking still remains an important study skill to be developed for all students if possible. Note-taking during lessons and when obtaining information from reference books or Internet sources is actually a very complex task; even very capable students can find the process difficult. The two main problems associated with taking notes are trying to identify the key points that are worth writing down, and (during an oral presentation) trying to do two things at once — listen to new information while still writing previous information. Some students, particularly those with poor fine-motor coordination, a physical disability or a learning disability, may be extremely slow when writing by hand, and thus get left far behind. Teachers must decide whether it is worth requiring such students to take notes at all, or whether it is more effective to provide a brief printed summary of the main points from the lesson for the student to use.

At secondary school level, it is often of great benefit to all students to distribute clear and concise study notes for each lesson, with main points summarised, key terms defined and assignment or homework tasks explicitly described. Another common approach is to encourage students with learning difficulties to borrow and copy the notes of other students after the lesson has ended. Teachers might also consider facilitating the process of note-taking by saying, at appropriate points in the lesson, 'This is worth writing down in your books' or simply saying directly, 'OK. Write this down'. It is also helpful to make frequent use of the whiteboard to summarise main points or conclusions; and to pause sometimes to ask certain students to read out what they have noted so far. If necessary, suggest what else might be written down and allow time for students to catch up. It is also worth the effort to talk with the class about note-taking as an essential study skill. How do good note-takers go about the task? What needs to be noted and what can be ignored? How do notes help you to study or do your assignments and homework?

The main barrier to learning for some students will continue to be the level of difficulty of the textbooks and other print materials they are required to read. For example, in both history (Harniss et al., 2001) and geography (Jitendra et al., 2001) textbooks are found to be poor resources when used by students with literacy difficulties. A strategy that teachers can use to help students read more confidently (including those reading in English as a second language) is to adjust the readability level of the material downwards (Conway, 1996; Janney & Snell, 2004). By simplifying text in terms of vocabulary load and syntactical complexity, or by augmenting text with additional cues and supports, teachers will assist students to develop fluency. As indicated in Chapter 5, studies have shown that improved fluency facilitates comprehension by allowing readers to devote cognitive effort to higher order processes of analysis and reflection, rather than to lower order word recognition and decoding (Pikulski & Chard, 2005). Standard curricular resources often need to be modified for students with learning difficulties, but such adaptations are very demanding of teachers' preparation time and ingenuity — so often they do not get done (Westwood, 2002; Yuen, Westwood & Wong, 2005).

Readability of print materials is not the sole determinant of students' level of understanding in social studies and related subjects. Other aspects of language competence are equally important. In particular, a student's grasp of vocabulary will influence his or her immediate understanding of information given orally by the teacher or presented via multimedia resources. Confusion quickly arises when instructions are not understood or when new terminology is introduced but not adequately explained and practised. Students with intellectual disability in particular can have major difficulties learning and remembering new vocabulary, or recalling the names of places and people (Waterfield, 2002).

Cognitive and affective development

In Chapter 7 on teaching science, attention was devoted to the way in which students' stages of cognitive development can limit to some degree the extent to which they deal easily with abstractions, comprehend symbolic representation and deduce cause–effect relationships. Although effective teaching can influence students' readiness to engage in new learning, it must be acknowledged that the same cognitive limitations do apply in social studies, history, geography and environmental education. Students will understand and learn most easily the information that comes from direct experience and links smoothly with what they already know and use (von Heyking, 2004). Indeed, the more removed from students' personal experience topics are, the more difficult they will be for the students to relate to and comprehend. In the domain of history, for example, Martin and Gummett (2001) remind teachers that new information must be linked in some way with students' own experiences or it will remain as isolated blobs of useless data soon forgotten. In social studies, Leming (2003) is cautious about the use of topics and contentious issues that require students to exercise high levels of critical thinking and to make judgments about events or policies of which they have no first-hand experience. Such tasks are often developmentally inappropriate.

It is not until students are in Year 5 of school that they begin to appreciate historical time and become capable of sequencing events into particular periods (Barton & Levstik, 1996; Ormrod, 2000). This is certainly not an argument for excluding history and 'historical thinking' from the curriculum for younger children or for students with learning difficulties. In the early years, however, history topics should relate mainly to a child's own personal and family 'history' and to local events.

An important part of 'thinking historically' involves the affective elements of identification, empathy and moral judgment — for example, being able to put yourself in the shoes of someone living at a different time, feeling what they would feel, and reacting in ways they would have reacted to the pressures and opportunities in their lives. What we know of children's development in the affective (feelings and values) domain suggests the young students and older students with learning difficulties or with intellectual disability may be significantly delayed in reaching a mature stage of moral reasoning and in appreciating conventional values (Piaget, 1965; Kohlberg, 1978). They may have difficulty putting themselves mentally in the situation of someone else and understanding how that person might feel or act. Piaget implied that normal children below the age of 11 or 12 years have difficulty judging the rightness or wrongness of motives that lead individuals to act in certain ways. Kohlberg believed that genuine interest in and compassion for others is not achieved until very much later (and in some cases is not even achieved in adulthood). Clearly, some of the intended affective outcomes from history and civics teaching are difficult to achieve for students whose development of empathy is delayed. However, Kohlberg believed that children can be helped to move to a higher level of understanding by encountering new views that challenge their own beliefs and stimulate them to form better arguments (Crain, 1992). There is a role for history and civics teaching in fostering such development.

In the teaching of geography and environmental education, as with history, it is important to match the students' cognitive level. It is essential to work from the real world, progressing gradually toward abstractions and principles. For example, some students with learning disabilities have problems using atlases and maps and in reading graphs (Waterfield, 2002). This is partly because 'proportional reasoning' does not emerge until adolescence, and without such understanding students have difficulty in dealing with scale. It is essential that students first gain experience in drawing plans and simple sketch-maps based on their own observations of the real environment. Later they can begin to appreciate scale and relative position by studying aerial photographs and relating these to sketch maps before being introduced to scale maps with conventional signs and grid reference systems.

Videos, computer simulations, photographs, models and field trips all help to make data for history, geography and environmental education more visual and concrete for all learners; such resources are vital for gaining and maintaining the interest of students with learning difficulties. Film and video are wonderful resources for introducing contrasting cultures, locations, climates and lifestyles. For students operating cognitively at a concrete level, film and video provide an experience of other worlds and other times. Similarly, video and film are acceptable substitutes for some environmental education field trips, particularly when the theme is, for example, logging in rainforests, but the school is in a city area.

Numeracy problems

Numeracy skills required in social studies and related subjects are only slightly less than those required for studying science. For example, graphs and charts depicting data such as annual rainfall figures, imports and exports, population growth or decline, increases or decreases in air pollution, death rate from specific diseases, comparison of family income in 1950s with income in 2006 and so forth often have to be read and interpreted. During field trips and other practical activities, new data often has to be recorded and charted, calculations have to be made and quantitative problems solved. Students with weak numeracy skills are at risk of failure in such situations.

One particular obstacle in relation to the study of history is the underdeveloped awareness that many children have of 'historical time'. This is not surprising because, compared with older persons, they have had much less direct experience of the passage of time and the changes it brings (von Heyking, 2004). The problem is that when young children or older students with learning difficulties study specific events in history, they may not place them in any meaningful order or context, and may for example believe that Joan of Arc lived at the time of the Roman Empire or that dinosaurs became extinct only shortly before the first aircraft was invented. Indeed, history has been a rich source of many well-known 'schoolboy howlers' due to such misconceptions (excuse the sexist term 'schoolboy'; I am sure schoolgirls have also generated many howlers, for which they should be given due credit). The message is that teachers need to anticipate problems that children naturally have in appreciating time sequences and contiguity in past events.

Problems associated with field trips

Field trips represent a very valuable component in active teaching and learning in geography, history and environmental education. Field trips motivate learners and bring situations to life in a way that no textbook or video can ever equal. Observing and learning in the real world is an ideal approach, as discussed in Chapter 3. Social benefits can also emerge from the cooperative learning that is usually a part of field work conducted off the school campus. However, field trips are not always planned and conducted effectively, so the desired learning outcomes are not achieved.

Students must have clearly defined objectives for a field trip, and must have specific activities and roles to carry out that will later contribute information to the consolidation stage. Efficient use of the available time is always important, and 'dead spots' should be avoided. Badly organised and managed field trips, where students are not clear concerning the purpose and their own responsibilities, result in poor learning.

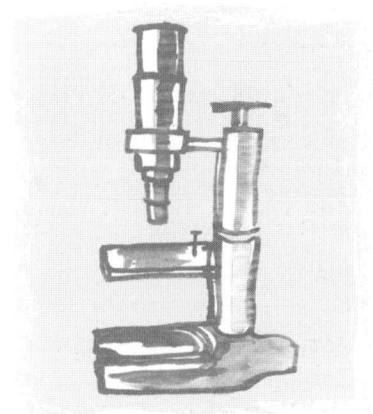

There are particular challenges for teachers when students with physical disabilities, or emotional and behavioural disorders, are to participate in field trips. The main issues relate to safety and behaviour management. Principles of inclusive education suggest that these students have a right (and perhaps an even greater need) to join in these excursions. The teacher may need to solicit advice from special education staff on appropriate strategies for taking these students out of school. They also need to ensure that they have additional adults available to assist with supervision and learning support.

Although in this chapter potential problems of learning in social studies and related subjects have been identified, it is important to end by reminding teachers that most students do enjoy working in these areas of the curriculum. The opportunities these subjects create for practical activity in the real world makes them ideal for stimulating students' interest, enhancing their understanding of everyday issues, extending their basic academic and communication skills, and building their confidence as learners.

Useful resources

History

Harris, R.J. & Luff, I. (2004). *Meeting special needs in history.* London: Fulton.

US Department of Education, Office of Intergovernmental and Interagency Affairs (2004). *Helping Your Child Learn History* (2nd edn). Washington DC: US Department of Education. Viewed 3 February 2006, (www.ed.gov/parents/academic/help/history/index.html).

It is not particularly easy to teach history in a meaningful way to students with intellectual disability or to those with very serious learning problems. However, these two resources provide some sound principles to guide teachers working in special schools or inclusive settings in mainstream schools.

Geography and environmental education

Heimlich, J.E. & Daudi, S.S. (1996). Environmental education and learners with disabilities. *EETAP Resource Library*, 1, April. Viewed 3 February 2006, (http://eelink.net/eetap/info1.pdf).

Higher Education Academy Subject Centre for Geography, Earth and Environmental Sciences (2006). Plymouth, UK: Higher Education Academy. Viewed 3 February 2006, (http://www.gees.ac.uk).

Staffordshire Learning Net (2005). *Special Needs Geography.* Stafford, UK: QLS/Staffordshire County Council. Viewed 3 February 2006, (http://www.sln.org.uk/geography/segsmain.htm).

Swift, D. (ed.) (2005). *Meeting SEN in the curriculum: Geography.* London: Fulton.

Geography and environmental education are two domains of study that provide abundant opportunities for success for students with learning difficulties. The subjects lend themselves to experiential learning, and present many opportunities to apply and extend language, literacy, numeracy and social skills. These resources will be of interest to teachers.

9 Support for learning

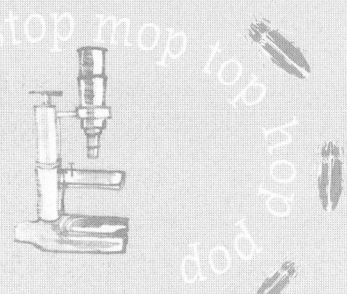

[In recent years] there have been a number of changes, both in the way that support for pupils with special educational needs has been organized and in the role of the 'specialist' teacher. Increasingly, specialist teachers have been encouraged to move away from a narrowly conceived role as 'remedial' teachers to one which has encompassed a much wider management brief, coordinating the overall response of the school to those pupils experiencing barriers to learning (Crowther, Dyson & Millward, 2001, p.85).

Students with learning difficulties usually require some form of additional support regardless of whether they are in mainstream classrooms or in special classes. The support may be necessary to facilitate their learning within the core curriculum, or it may be designed to assist them to overcome emotional, social or behaviour difficulties. Sometimes the support comes in human form, and sometimes as differentiated or specially selected methods and materials (Heeks & Kinnell, 1997). In the case of students with significant disabilities, the support may be needed to help them acquire skills that other students learn without assistance, such as mobility, self-care and communication (Foreman, Bourke & Mishra, 2001). Support may be provided not only through school-based resources but also through access to additional specialist services and resources beyond those available in the school, including psychological assessment, speech therapy and counselling (Dettmer, Thurston & Dyck, 2005).

It is not only students who require support; their teachers also need personal and professional support in order to cope with the ever-increasing demands of their job, and to assist students with learning or adjustment problems. Teachers need help in developing further their own professional skills. Without adequate support, teachers often develop or maintain very negative attitudes toward students with learning difficulties, and may find the increase in workload and responsibility stressful (Forlin, 2000; 2001a). They may continue to prefer to hand over to the so-called 'experts' any student who does not manage to learn easily in their classrooms.

Providing support in schools

The use of additional human resources has been one of the main ways in which education systems in different countries have responded to students with special needs in mainstream schools. Many countries now employ 'support teachers' or 'resource teachers', and may also supplement the effort of these teachers by using paraprofessionals — variously known as 'classroom assistants', 'learning support assistants' (LSAs), 'integration aides', 'teacher aides' or 'student support officers'. These additional personnel, together with a growing number of volunteer helpers in schools, are regarded as integral and essential components of an effective support system for students integrated into inclusive classrooms (Fox, 2003). In recent years, resource teachers have been encouraged to widen the scope of their remedial help so that they work more in support of other teachers in the school and less in direct teaching of students with learning difficulties in a withdrawal room. Co-teaching and in-class support are increasingly presented as desirable models of service delivery in inclusive schools.

In Britain, the presence of the 'special educational needs coordinator' (SENCO) in each school facilitates the deployment and coordination of these human resources to provide the optimum school-based support system. In other countries (e.g. Australia, New Zealand), the coordination and management of the support system is more likely to be in the hands of a 'special needs team' or ' learning support committee' within the school.

It is vital that any school-based support system should be well-managed, with all teachers and paraprofessionals fully aware of their roles and responsibilities, and parents aware of the scope and purpose of services offered. The school should have a written policy indicating the manner in which human and material resources are to be used for support, and identifying roles and responsibilities.

Support teachers

Support teachers are to be found in the school systems of many different countries. They may have different titles but they all carry out a fairly similar role of assisting students with special needs and their teachers (Emanuelsson, 2001). Some support teachers are based in a single school, with responsibilities only to the students and teachers in that school. Others may work on a regional or visiting service basis, and provide part-time assistance to several different schools.

Originally, the role of the support (remedial) teacher was to provide direct assistance to students with disabilities or learning difficulties, either through in-class support or more commonly by withdrawing students for intensive tuition in small groups. While this individualised and intensive teaching certainly did help some students make better progress, it also resulted in the support teacher's expertise being available only to a very limited number of students. The students were usually handed over by the class teacher to the support teacher — and, having passed the problem over to someone else, the mainstream class teachers were under no pressure to develop additional skills themselves to meet the needs of these students. Often the support offered to the students was some variation of fairly traditional 'remedial' teaching, having few if any links with the mainstream program. Studies had shown that this type of teaching might raise a student's basic skills while the support was being provided, but any gains the student made quickly disappeared when he or she returned to the mainstream classroom full time. In addition, withdrawal from class was not liked by the students concerned because it identified them in the peer group as being 'different' in some way, and also tended to fragment their learning experiences in the mainstream program.

Forlin (2001b) indicates that over the past 20 years the role of the support teacher has changed very significantly from serving a limited number of students directly to serving a much larger pool of students indirectly through supporting classroom teachers. Under this model the classroom teacher is helped to acquire some new teaching strategies and becomes better able to adapt the mainstream curriculum where necessary for a wider range of students. The support teacher may, for example, pass on to the mainstream teacher many of the strategies for differentiation described in Chapter 1. The class teacher is also given assistance in developing or locating appropriate alternative resources for individual students, and suggestions may be made on grouping the class more effectively to allow the teacher to give certain students individual attention during a lesson. To facilitate this model, the support teacher has had to become both a direct helper and an adviser to mainstream teachers. Accomplishing this role of 'helper–adviser' requires support teachers to use the process now known in most countries as *collaborative consultation* (Lacey, 2001). Collaborative consultation and teamwork are considered essential ingredients for effective support. The collaborative model is discussed more fully towards the end of this chapter.

The typical duties of a support or resource teacher include:

- Providing advice and assistance to other teachers in devising inclusive curricula and differentiated materials.
- Co-teaching with regular class teachers.
- Collaboration with parents, teachers and other professionals in the development and implementation of individual education plans (IEPs).
- Helping schools establish their own internal support systems.
- Liaising with outside agencies and services.
- Regular evaluation of progress made by students receiving various forms of additional support.
- Providing occasional in-class support to students with special needs.
- Occasional withdrawal of groups or individual students for intensive teaching.
- Educational assessment of students.
- Diagnosis of particular areas of difficulty in learning.
- Planning behaviour management interventions.
- Providing school-based and regional in-service training workshops for teachers.

Studies indicate that mainstream teachers value the passing on of expertise, the occasional withdrawal of students for group teaching and the production of differentiated materials. In some schools, however, support teachers are still used quite extensively for the older style of withdrawal teaching, even though the model favoured under inclusive schooling philosophy is for students to remain in their own classes for support. Some research indicates that a combination of in-class support together with regular withdrawal for intensive instruction produces the best gains in achievement (Marston, 1996).

In-class support

Co-teaching, with the support teacher and the class teacher working together, is one approach with potential to increase flexibility in the amount of time and attention available to individual students during lessons. Simultaneously it capitalises on teachers' combined expertise and facilitates the sharing of teaching or organisational strategies. Co-teaching is frequently recommended as one valuable way of helping classroom programs become more inclusive (e.g. Austin, 2001). Typical models of co-teaching are not restricted simply to support teachers working with mainstream teachers; co-teaching can also operate very successfully when two or more mainstream teachers decide to work collaboratively.

The purposes of co-teaching with a support teacher are:

- Sharing responsibility for planning and implementing differentiated lessons.
- Enhancing opportunities for differentiated group work.
- Organising effective in-class support linked appropriately with the content of the general curriculum.
- Helping the regular class teacher or subject specialist develop new skills in adapting resources, modifying tasks and managing time.

Obstacles to co-teaching include the unwillingness of some teachers to allow another teacher to be present in the classroom, lack of time to sit down with one's co-teacher to plan lessons that will utilise the presence of two teachers to best advantage, and clashes in teaching style or management techniques (Lacey, 2001). There are some signs that resistance to having another teacher in the classroom is gradually diminishing, perhaps because various forms of teaming have become more commonplace in schools in recent years (Fletcher-Campbell & Cullen, 2000), but lack of time to plan together, and incompatible teaching styles or personalities, remain problematic. Nevertheless, co-teaching does offer the possibility of matching instruction more closely to the students' capabilities, and of increasing the amount of assistance available to individual students during a lesson. For these reasons schools should be encouraged to consider this option whenever human resources permit.

Paraprofessionals

In the past decade the value of paraprofessionals assisting within the classroom has been recognised in most education systems (DfES, 2004; Wallace et al., 2001). Dettmer, Thurston and Dyck (2005) suggest that paraprofessionals should be given the status of 'contributing partners' in the educational process, not simply persons who carry out mundane tasks under the direction of the teacher. They need to be included as essential members of the school's support team and encouraged to contribute ideas to enhance classroom programs as well as improve the quality of support. Communication channels within the school should always ensure that assistants are kept fully in touch with school-based issues. The valuable work done by paraprofessionals needs to be openly acknowledged by the school whenever possible (Giangreco, Edelman & Broer, 2001).

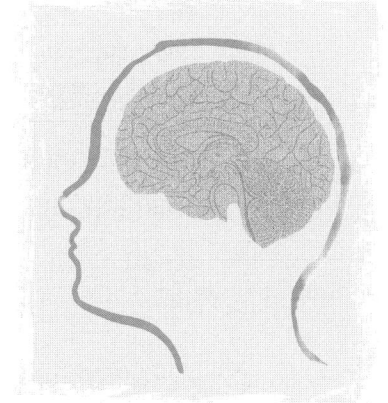

According to Simon (2001), it is the professionalism of typical classroom assistants that makes them such valuable resources in any school. It might be better to conceptualise their role as that of 'assistant teacher', rather than 'teacher's assistant', thus emphasising the fact they are indeed an integral component in the teaching and learning process. At all times assistant teachers should be clear about their duties and should not have doubts about their own status in the school or classroom. Their roles should made explicit within the school's own written policy on support for learning.

The assistant always works under the guidance of the teacher. While assistants are not responsible for determining the details of the curriculum content to be followed by a student, or for setting the objectives and selecting methods, they can be instrumental in helping the student access the curriculum and achieve objectives. It will always remain the responsibility of the teacher, acting on information from all relevant sources, to determine the details of the student's program.

Simply stated, a classroom assistant can contribute to the teaching and learning process in at least the following ways:

- Working with individual students.
- Working with small groups.
- Providing an extra pair of hands in the classroom generally.
- Modifying and developing materials.

More specifically, the types of support provided by classroom assistants include:

- Showing positive interest in a student's work.
- Listening to any problems and worries a student may have.
- Providing praise and encouragement.

- Keeping a student on task.
- Explaining or interpreting the teacher's instructions again to certain students.
- Helping to interpret printed materials or blackboard notes.
- Acting as a scribe for a student who has difficulty in writing.
- Preparing and adapting materials.
- Correcting work and providing feedback.
- Partnering a student in a computer-assisted learning activity.
- Assisting with group work activities.
- Helping implement an individual education plan (IEP).
- Supporting learning within literacy and numeracy programs.
- Listening to students read.
- Reading to students.
- Helping to establish and manage a paired reading program.
- Managing behaviour.
- Assisting with administrative duties and paperwork.
- Supporting students with disabilities when they are carrying out practical tasks.
- Acting as reader, interpreter and prompter for students with hearing or communication difficulties.

Classroom assistants have an important role in pastoral care. They can be invaluable as an extra 'ear' for detecting students' worries and disputes. Paraprofessionals can fulfil the important role of trusted adults who have time to attend to individual students' concerns and interests. When possible, the assistant should be encouraged to provide support to *all students* in the class, not just those who have been identified as having special needs. The presence of an assistant in the classroom should enable the teacher to give increased attention when necessary to individual students.

If paraprofessionals are to provide an optimum service in schools, they require high-quality training for the duties they undertake (DfES, 2004; Farrell, Balshaw & Polat, 2000). Teachers, too, may need guidance in how best to manage these other adults in the classroom and how to work collaboratively with them. This has not been part of most teachers' initial professional training, so the help has to come mainly from in-service programs.

As indicated above, the use of classroom assistants can be highly beneficial, especially in the support of students with significant disabilities in regular classrooms. It must be mentioned, however, that problems can arise and obstacles may exists to the smooth operation of paraprofessional support. For example, paraprofessionals may not fully recognise that the support they provide should help a student develop independence. This will not be achieved if the assistant helps the student too much, or simplifies tasks to a level where there is no challenge. During their training, paraprofessionals are taught to encourage independent effort and autonomy in the students. They are also advised not to allow students to become lazy and overprotected, expecting to have too much done for them. Tennant (2001) suggests that students with disabilities who are constantly accompanied by a support assistant throughout the day may find it very difficult to make friends with other children.

Some of the difficulties already identified when support teachers and mainstream teachers try to work together apply equally to paraprofessionals and teachers. There can be a clash of personalities or styles, there may be a lack of sufficient communication between teacher and assistant, and shortage of time can lead to inadequate shared planning. In recent years there has been a rapid increase in the number of valuable in-service courses and books available to help teachers and paraprofessionals work effectively together (e.g. Balshaw, 1999; Fox, 2003).

Volunteers

Some countries have made increasing use of volunteer helpers in the classroom. This applies in both mainstream and special schools. For example, in many parts of Australia schools create what are often called *learning assistance programs* (LAPs) using parents, grandparents, retired teachers, student teachers or other unpaid workers to help students with learning problems — and sometimes to work with gifted and talented students as part of a 'mentor' scheme.

When assisting students with learning problems, the roles given to such helpers include:

- Listening to individual students read (providing extra practice).
- Helping students with their writing and spelling.
- Helping students check or prepare homework assignments.
- Playing learning games with a small group.
- Helping students with physical disabilities participate in activities.
- Making resources.
- Sitting with a student to keep him or her on task.
- Working with a group of students to help them complete tasks set by the teacher.

These volunteers perform a very valuable service in schools. The amount of individual attention they can devote to students with special needs is far greater than most teachers can afford to give. Obviously their roles are very similar to those already discussed for paraprofessional helpers; the main difference is that these helpers are not paid.

It is essential to identify *reliable* helpers. Major problems will arise if volunteers arrive late, or do not come on the days they are expected in the school. From the start they must always know that they work under the direction of the teacher and must not take it into their own hands to set work for the students or to do things in a different way from what the teacher had planned. They must also be very clear about issues of discipline and what they should do (and not do) if a student misbehaves.

When selecting volunteers for support work in school it is essential to stress:

- Punctuality.
- Reliability.
- Responsibility.

The Special Educational Needs Coordinator (SENCO)

In Britain, under the original and the revised *Special Educational Needs Code of Practice* (DfES, 2001) every school is required to have a member of staff identified as the Special Educational Needs Coordinator (SENCO). The revised *Code of Practice* came into force in 2002. The duties associated with the SENCO position have been made explicit within the *Code of Practice* and also in a series of guidelines. Additional information on standards and competencies for SENCOs and specialist teachers was prepared by the Teacher Training Agency (TTA) in 1998, 1999 and 2000.

The key responsibilities of the SENCO, as specified in the revised *Code of Practice*, include the following:

- Overseeing the day-to-day operation of the school's SEN policy.
- Coordinating provision for students with special educational needs.
- Liaising with and advising fellow teachers.
- Managing learning support assistants.
- Overseeing the records on all students with special educational needs.
- Liaising with parents of students with special educational needs.
- Contributing to the in-service training of staff.
- Liaising with external agencies including the LEA's [local education authority's] support and educational psychology services, health and social services and voluntary bodies.
  (DfES, 2001, para. 5:32, p.50)

According to Shuttleworth (2000), a key role of the SENCO is to encourage higher student success rates by influencing colleagues to adjust the curriculum and methods. Increasingly, the role of a SENCO has come to embrace 'whole-school staff development', with the aim of helping teachers adopt more inclusive practices and cope with the demands of mixed-ability teaching. It is stated in *National Standards for Special Educational Needs Coordinators* (TTA, 1998, p.5) that:

> The SENCO's fundamental role is to support the head teacher in ensuring that all staff recognise the importance of planning their lessons in ways that will encourage the participation and learning of all pupils [and] the SENCO plays a key role in supporting, guiding, and motivating colleagues, particularly in disseminating examples of effective practice in relation to pupils with special educational needs.

It is not easy to find one person who possesses all the qualities necessary to carry out the role of SENCO. Cheminais (2001) suggests that a SENCO will need to have a proven record of successful teaching, and possess interpersonal skills necessary for motivating and inspiring staff and students. Ideally, SENCOs also need to have skill in monitoring improvements in teaching and learning, proven ability to lead and manage a team, ability to implement change effectively, a personal commitment to school improvement and proficiency in using ICT.

The role of a SENCO must be one of the most difficult ever created for a teacher, particularly in a large school comprising many students with learning and behavioural problems (Crowther, Dyson & Millward, 2001). The duties expected of the SENCO have caused a major reorientation in working practices and a significant increase in workload (Davies, Garner & Lee, 1998). Although many reports and articles put a positive spin on how SENCOs in general are managing their very challenging task (e.g. HMI, 1997), there is still much evidence to indicate that the role is problematic (Dyson & Millward, 2000). In particular, the 'change agent' and staff development roles seem difficult to fulfil, although some individual schools have progressed well in this direction (Cowne, 2003). SENCOs need support, particularly from the senior management group in the school.

Schools in other countries often have a member of staff who carries out many of the duties identified above, although he or she may have a somewhat different title. It has also become fairly common practice for schools to establish a team of teachers who are responsible for coordinating provision of support for learning.

Teacher assistance teams (TATs)

The emphasis in responding to special educational needs and learning difficulties has quite clearly moved toward a 'whole-school approach'. The provision for students with special needs has to be seen as an integral part of planning for all students. A view developed during the 1990s that schools generally possess a significant pool of professional experience and expertise that had not been tapped effectively in the past. This school-based expertise could be utilised to help schools become more self-sufficient in providing support for their own students and teachers. Self-sufficiency has also been forced on many schools by the reality that external resources and services are finite and demand always seems to outstrip supply.

To facilitate school-level responses to the needs of individual students and their teachers, one system that appears to operate well is the establishment of a 'teacher assistance team' (TAT) or 'special needs team' (Bradley & Roaf, 1995; Creese, Daniels & Norwich, 1997). The team is made up from teachers who have a broad knowledge of different teaching strategies and are conversant with flexible ways of managing classrooms. Any teacher in the school who has a student with learning or behavioural difficulties is encouraged to join the team, and through open discussion with colleagues, strategies are developed to meet these students' needs. Many classroom problems can be resolved and teacher-support structures put in place by drawing upon the existing pool of knowledge and experience.

Referral of a problem to a school-based team usually brings far more rapid action and assistance than any referral to outside agencies could ever achieve. The system is also said to help individual teachers become more confident in sharing problems with colleagues, and eventually more self-sufficient in dealing with students' difficulties and differences.

Teachers' networks for support can extend beyond the school itself. It is extremely helpful for staff with similar concerns across a number of different schools to be able to contact one another to discuss problems and devise possible solutions or develop resources. Sometimes these networks become established informally, but it should also be seen as part of a visiting support teacher's role to help establish contact between teachers with similar professional needs in different schools.

Collaborative consultation

Almost every supporting role described in this chapter involves a process of collaboration with others. Such collaboration is considered a key to effective learning support. In order to implement the 'whole-school approach' to special educational needs, it is necessary to ensure a high level of collaboration and cooperation among staff in schools, and between schools and outside agencies. Establishing a positive working relationship for collaboration among teachers and other professionals must involve mutual respect, trust and gradual adaptation to one another's style of working. Collaboration only works successfully if the contributions of all parties are valued.

Consultation involves seeking advice from others who possess appropriate knowledge and experience in order to solve a problem or meet a personal or professional need. Teachers frequently find it necessary to seek such advice when working with students who have learning difficulties. They may need advice from colleagues, advisory teachers, parents, educational psychologists, counsellors and outside professionals. In the context of learning difficulties, Dettmer, Thurston and Dyck (2005) describe collaborative school consultation as a process in which school personnel and families confer, consult and collaborate as a team to identify a student's learning and behavioural needs, and to plan, implement, evaluate and revise the program that is expected to serve those needs.

The meetings required to design a student's individual education plan (IEP) should be good examples of collaborative consultation. Each member contributes information, seeks additional advice and suggests possible solutions for the student's problems. The end product is a combination of ideas translated into a relevant plan of action. No single expert is likely to have all the knowledge or skills required to address the special needs and difficulties experienced by some students, but by sharing ideas, knowledge and support strategies, a teacher can be helped to implement a more effective program. The long-term aim of any collaborative consultation approach is to share expertise, help all teachers become better able to meet their students' individual needs, and assist schools to develop their own internal systems of support.

Collaborative consultation also occurs when a teacher or a school principal approaches someone who has experience or professional knowledge in a particular field for advice. For example, the principal may seek the help of a behaviour management team if a particular class in the school is troublesome, or if a specific student is out of control. The team may suggest that the student's parents and an educational psychologist be involved in the discussions. They collaborate by examining the problem in detail and formulating an agreed plan of action. After the plan is implemented, they help to monitor its effectiveness.

Although collaborative consultation may occur over any particular matter or problem that cannot be easily solved by the teacher alone, the following tend to be the most common areas where teachers seek additional guidance:

- Interpreting the results from any formal assessment of a student with a learning difficulty.
- Obtaining background information on a particular form of disability or impairment, and the impact of that disability on the student's learning and adjustment.
- Setting priorities within the objectives contained in a student's special program.
- Selecting appropriate teaching methods.
- Creating or adapting resource materials and equipment (e.g. for a student with a physical disability).
- Seeking advice and assistance in setting up a behaviour modification program.
- Arranging additional support.
- Making links with outside agencies.
- Creating a support network for the student and for the teacher.
- Providing ongoing and follow-up evaluations of a student's progress.
- Extending support to the home environment.

Although collaboration and consultation are regarded as essential processes in effective support, there are often problems in engaging in them effectively. For example, a study comparing collaborative consultation practices in two states of Australia revealed that the major obstacles encountered by teachers were lack of time to plan together with others, lack of time to implement plans that had been drawn up, lack of training in specific skills and processes needed for teamwork, and inadequate back-up from outside services and agencies (Westwood & Graham, 2000). Emanuelsson (2001) identified a common problem across most countries, namely that support systems and support teachers themselves tend to be reactive rather than proactive when providing support. For example, they are more likely to be called in by a school after a serious problem has arisen, rather than consulted by school staff when they are first establishing support systems.

Effective teaching, as described throughout this book, will help reduce the number of students experiencing learning difficulties in school, but even effective teaching will not eliminate all problems. The information provided in this chapter may help schools establish support systems that address the urgent needs of students who continue to have difficulties in learning.

Useful resources

Support teachers

European Journal of Special Needs Education (2001). *16*, *2*, June.
A special issue on the role of support teachers.

Teacher assistance teams (TATs)

Creese, A., Daniels, H. & Norwich, B. (1997). *Teacher support teams in primary and secondary schools.* London: Fulton.
This book offers some very practical advice and resource materials for setting up teacher assistance teams in schools.

Collaborative consultation

Dettmer, P., Thurston, L. & Dyck, N. (2005). *Consultation, collaboration and teamwork for students with special needs* (5th edn). Boston: Pearson Allyn & Bacon.
Contains useful information on all aspects of collaborative consultation.

General

Cowne, E. (2003). *The SENCO handbook: Working within a whole-school approach* (3rd edn). London: Fulton.
Fox, G. (2003). *A handbook for learning support assistants* (2nd edn). London: Fulton.
Morgan, J. & Ashbaker, B.Y. (2001). *A teacher's guide to working with paraeducators and other classroom aides.* Alexandria, VA: Association for Supervision and Curriculum Development.
These titles are useful for a range of topics discussed in this chapter.

References

AAAS (American Association for the Advancement of Science) (1990). *Science for all Americans (Project 2061)*. New York: Oxford University Press.

AAAS (American Association for the Advancement of Science) (1993). *Benchmarks for science literacy*. New York: Oxford University Press. Viewed 6 February 2006, (http://www.project2061.org/tools/benchol/bchin.htm)

Adams, G. & Engelmann, S. (1996). *Research on direct instruction: 20 years beyond DISTAR*. Seattle: Educational Achievement Systems.

Adams, M.J. (1990). *Beginning to read: Thinking and learning about print*. Cambridge, MA: MIT Press.

Adventures of Jasper Woodbury (1992). Set of 12 interactive videodiscs. Mahwah, NJ: Learning Inc.

Alexander, R.J. (1995). *Versions of primary education*. London: Routledge.

Alexiades, J., Gipson, S. & Morey-Nase, G. (2001). Deconstructing 'the classroom': Situating learning with the help of the World Wide Web. In A. Herrmann and M.M. Kulski (eds) *Expanding horizons in teaching and learning: Proceedings of the 10th Annual Teaching Learning Forum*. Perth: Curtin University of Technology. Viewed 6 February 2006, (http://lsn.curtin.edu.au/tlf/tlf2001/alexiades.html).

Arreaga-Mayer, C. (1998). Increasing active student responding and improving academic performance through classwide peer tutoring. *Intervention in School and Clinic, 34, 2*, 89–95.

Ashlock, R.B. (1998). *Error patterns in computation* (7th edn). Upper Saddle River, NJ: Merrill.

ASE (Association for Science Education) (1997). *Access to science education: Policy*. Hatfield, UK: ASE.

Austin, V.L. (2001). Teachers' beliefs about co-teaching. *Remedial and Special Education, 22, 4*, 245–255.

Australian School Library Association and Australian Library and Information Association (2001). *Learning for the future: Developing information services in Australian schools* (2nd edn). Melbourne: Curriculum Corporation.

Ausubel, D.P. (1968). *Educational psychology: A cognitive view*. New York: Holt, Rinehart & Winston.

Bahar, M., Johnstone, A.H. & Hansell, M.H. (1999). Revisiting learning difficulties in biology. *Journal of Biological Education, 33, 2*, 84–86.

Balshaw, M. (1999). *Help in the classroom*. London: Fulton.

Bancroft, J. (2002). A methodology for developing science teaching materials for pupils with learning difficulties. *Support for Learning, 17, 4*, 168–175.

Barkley, R.A. (2003). Attention-Deficit/Hyperactivity Disorder. In E.J. Mash & R.A. Barkley (eds) *Child psychopathology* (2nd edn, pp.75–143). New York: Guilford Press.

Barr, R., Blachowicz, C., Katz, C. & Kaufman, B. (2002). *Reading diagnosis for teachers* (4th edn). Boston: Allyn & Bacon.

Barton, K.C. & Levstik, L.S. (1996). 'Back when God was around and everything': Elementary children's understanding of historical time. *American Educational Research Journal, 33*, 419–454.

Barry, K. (1995). Lecturing, explaining and small-group strategies. In Maltby, F., Gage, N. & Berliner, D. (eds) *Educational psychology: An Australian and New Zealand perspective* (pp.356–417). Brisbane: Wiley.

Batten, M., Marland, P. & Khamis, M. (1993). *Knowing how to teach well.* Hawthorn, Vic: Australian Council for Educational Research.

Bay, M., Staver, J., Bryan, T. & Hale, J. (1992). Science instruction for the mildly handicapped: Direct instruction versus discovery teaching. *Journal of Research in Science Teaching, 29,* 555–570.

Bell, D. (1999). *Accessing science in the primary school: Meeting the challenges of children with learning difficulties.* Paper presented to the AARE-NZARE Conference, Melbourne.

Bell, D. (2002). Making science inclusive: Providing effective learning opportunities for children with learning difficulties. *Support for Learning, 17, 4,* 156–161.

Bereiter, C. & Scardamalia, M. (1987). *The psychology of written composition.* Hillsdale, NJ: Erlbaum.

Binder, C. & Watkins, C. (1990). Precision teaching and direct instruction: Measurably superior instructional technology in schools. *Performance Improvement Quarterly, 3, 4,* 74–96.

Birsh, J.R. (2005). *Multisensory teaching of basic language skills* (2nd edn). Baltimore: Brookes.

Block, J. (1971). *Mastery learning: Theory and practice.* New York: Holt, Rinehart & Winston.

Bloom, B.S. (1971). *Mastery learning.* New York: Holt, Rinehart & Winston.

Bloom, B.S. (1984). The 2-Sigma problem: The search for methods of group instruction as effective as one-to-one tutoring. *Educational Researcher, 13, 6,* 4–16.

Boaler, J. (1997). *Experiencing school mathematics.* Buckingham: Open University Press.

Booker, G. (2004). Difficulties in mathematics: Errors, origins and implications. In B.A. Knight & W. Scott (eds) *Learning difficulties: Multiple perspectives* (pp.129–140). Frenchs Forest, NSW: Pearson.

Booth, W.R. (2000). *Anchored instruction: An introduction.* Educational Technology. Greely, CO: University of Northern Colorado. Viewed 6 February 2006, (http://www.coe.unco.edu/ET500/Booth/TSLD001.HTM).

Bowen, T. (2004). *Methodology challenge: Task-based learning.* Onestopenglish.com. London: Macmillan. Viewed 6 February 2006, (http://www.onestopenglish.com/ teacher_support/methodology/archive/teachingapproaches/task_based_ learningtb. htm).

Bradley, C. & Roaf, C. (1995). Meeting special educational needs in the secondary school: A team approach. *Support for Learning, 10, 2,* 93–99.

Brandt, R. (1992). Reconsidering our commitments. *Educational Leadership, 50, 2,* 5.

Bransford, J.D. & Stein, B.S. (1993). *The ideal problem solver* (2nd edn). New York: Freeman.

Brennan, S. & Robinson, G. (1998). Four approaches to comprehension instruction. *Australian Journal of Learning Disabilities, 3, 4,* 12–19.

Bristow, J., Cowley, P. & Daines, B. (1999). *Memory and learning: A practical guide for teachers.* London: Fulton.

Brophy, J. (1998). Failure syndrome students. *ERIC Digest EDO-PS-98-2.* ERIC Clearinghouse on Elementary and Early Childhood Education. Champaign, IL: University of Illinois.

Brophy, J. & Good, T. (1986). Teacher behavior and student achievement. In M. Wittrock (ed.) *Handbook of research on teaching* (3rd edn, pp.328–375). New York: Macmillan.

Brown, A.L. & Campione, J.C. (1994). Guided discovery in a community of learners. In K. McGilly (ed.) *Classroom lessons: Integrating cognitive theory and classroom practice.* Cambridge, MA: MIT Press.

Brown, J.S., Collins, A. & Duguid, S. (1989). Situated cognition and the culture of learning. *Educational Researcher, 18, 1,* 32–42.

Brown, K.J., Morris, D. & Fields, M. (2005). Intervention after Grade 1: Serving increased numbers of struggling readers effectively. *Journal of Literacy Research, 37, 1,* 61–94.

Brown, M. (2000). What kinds of teaching and what other factors accelerate primary pupils' progress in acquiring numeracy? In *Improving numeracy learning: Research conference 2000 proceedings* (pp.3–5). Melbourne: Australian Council for Educational Research. Viewed 6 February 2006, (http://www.acer.edu.au/workshops/documents/conference_proceedings2000.pdf).

Bruce, M.E. & Robinson, G. (2004). Clever kids: A metacognitive and reciprocal teaching program. *Australian Journal of Learning Disabilities, 9, 3*, 19–33.

Bruner, J. S. (1966). Some elements of discovery. In L.S. Shulman & E.R. Keislar (eds) *Learning by discovery: A critical appraisal* (pp.101–113). Chicago: Rand McNally.

Buck Institute for Education (2002). *Project based learning handbook.* Novato, CA: Buck Institute for Education. Viewed 1 February 2006, (http://www.bie.org/pbl/pblhandbook/intro.php).

Burden, P.R. & Byrd, D.M. (2003). *Methods for effective teaching* (3rd edn). Boston: Allyn & Bacon.

Burgess, J. (2003). Pay attention please! Strategies to reduce the learning difficulties experienced by children with attention and concentration problems. *Australian Journal of Learning Disabilities, 8, 3*, 8–14.

Burkill, S., Rawling, E., Bednarz, S. & Lidstone, J. (1999). *Developing an international network for learning and teaching geography in higher education: Reflections on experience in school education.* Cheltenham, UK: Geography Discipline Network/University of Gloucestershire. Viewed 6 February 2006, (http://www2.glos.ac.uk/gdn/ hawaii/ resef.htm).

Burns, M. (1998). Can I balance arithmetic instruction with real-life math? *Instructor, 107, 7*, 55–58.

Butler, F.M., Miller, S.P., Lee, K.H. & Pierce, T. (2001). Teaching mathematics to students with mild to moderate mental retardation: A review of the literature. *Mental Retardation, 39, 1*, 20–31.

Bygate, M., Skehan, P. & Swain, M. (eds) (2001). *Researching pedagogic tasks: second language learning, teaching, and testing.* Harlow, UK: Longman.

Campbell, L., Flageolle, P., Griffith, S. & Wojcik, C. (2002). Resource-based learning. In M. Orey (ed.) *Emerging perspectives on learning, teaching, and technology.* Online book. Athens, GA: University of Georgia College of Education. Viewed 1 February 2006, (http://www.coe.uga.edu/epltt/RBL.htm).

Carnine, D. (2000) *Why education experts resist effective practices (and what it would take to make education more like medicine).* Washington, DC: Thomas B. Fordham Foundation. Viewed 31 January 2006, (http://www.edexcellence.net/doc/carnine.pdf).

Carnine, D.W. (2004). *Direct instruction reading* (4th edn). Upper Saddle River, NJ: Pearson-Prentice Hall.

Carnine, D., Dixon, R. & Silbert, J. (1998). Effective strategies for teaching mathematics. In E. Kameenui & D. Carnine (eds) *Effective teaching strategies that accommodate diverse learners* (pp.93–112). Columbus, OH: Merrill.

Carroll, J. (1963). A model for school learning. *Teachers College Record, 64*, 723–733.

Carter, C.J. (2001). *Reciprocal teaching: Application of a reading improvement strategy on urban students in Highland Park.* Geneva: International Bureau of Education.

Cartwright, S. (2001). RBL revisited. *Access* (ASLA), 15, 2, 18.

Castles, A. (2005). Why phonics is not an F word. *Learning Difficulties Australia Bulletin, 37, 3*, 3.

Cavanagh, S. (2004). NCLB could alter science teaching. *Education Week, 24, 11*, 1–3.

Cawley, J.F. & Parmar, R.S. (2001). Literacy proficiency and science for students with learning disabilities. *Reading and Writing Quarterly, 17*, 105–125.

Chai, J.I. & Hannafin, M. (1995). Situated cognition and learning environments: Roles, structures and implications for design. *Educational Technology Research & Development* *43, 2*, 53–69.

Chalkley, B. & Waterfield, J. (2001). *Providing learning support for students with hidden disabilities and dyslexia undertaking fieldwork and related activities.* Cheltenham, UK: Geography Discipline Network (GDN). Viewed 6 February 2006, (http://www2.glos.ac.uk/gdn/disabil/hidden/hidden.pdf).

Chan, C.W.M., Chang, R.M.L., Westwood, P. & Yuen, M.T. (2002). Teaching adaptively: How easy is it in practice? A perspective from Hong Kong. *Asia-Pacific Educational Researcher, 11, 1*, 27–58.

Chan, L.K.S. & Dally, K. (2001). Instructional techniques and service delivery approaches for students with learning difficulties. *Australian Journal of Learning Disabilities, 6, 3*, 14–21.

Cheminais, R. (2001). *Developing inclusive school practice.* London: Fulton.

Churton, M.W., Cranston-Gingras, A. & Blair, T.R. (1998). *Teaching children with diverse abilities.* Boston: Allyn & Bacon.

Classroom Compass (1998), 1, 2, Fall issue. Viewed 1 February 2006, (http://www.sedl.org/scimath/compass/v01n02/welcome.html).

Clegg, J. (2005). CLIL: Content and Language Integrated Learning. *Children and Teenagers, 2*, 36–38.

Cole, P. & Chan, L. (1987). *Teaching principles and practices.* New York: Prentice Hall.

Cole, P. & Chan, L. (1990). *Methods and strategies for special education.* New York: Prentice Hall.

Coleman, M.R. (2001). Curriculum differentiation: Sophistication. *Gifted Child Today Magazine, 24, 2*, 24–25.

Collins, A., Brown, J.S. & Holum, A. (1991). Cognitive apprenticeship: Making thinking visible. *American Educator*, Winter issue. Viewed 1 February 2006, (http://www.21learn.org/arch/articles/brown_seely.html).

Collis, K.F. & Romberg, T.A. (1992). *Collis-Romberg mathematical profiles.* Melbourne: Australian Council for Educational Research.

Committee on Undergraduate Science Education (1997). *Science teaching reconsidered.* Washington, DC: National Academy Press.

Conway, J. (1997). *Educational technology's effect on models of instruction.* Newark, DE: Judith Conway/University of Delaware. Viewed 6 February 2006, (http://copland.udel.edu/~jconway/EDST666.htm).

Conway, R. (1996). Curriculum adaptations. In P. Foreman (ed.) *Integration and inclusion in action* (pp.145–190). Sydney: Harcourt Brace.

Cooper, J.D. & Kiger, N. (2006). *Literacy: Helping children construct meaning* (6th edn). Boston: Houghton Mifflin.

Corrie, L. (1995). Memory, cognitive processing and the transfer of learning. In F. Maltby, N. Gage & D. Berliner (eds) *Educational psychology: An Australian and New Zealand perspective* (pp.256–306). Brisbane: Wiley.

Cowne, E. (2003). *The SENCO handbook: Working within a whole-school approach* (3rd edn). London: Fulton.

Crain, W.C. (1992). *Theories of development* (3rd edn). Englewood Cliffs, NJ: Prentice Hall.

Creese, A., Daniels, H. & Norwich, B. (1997). *Teacher support teams in primary and secondary schools.* London: Fulton.

Croll, P. & Moses, D. (2000). *Special needs in the primary school: One in five?* London: Cassell.

Cromley, J.G. (2005). *Learning with computers: The theory behind the practice.* Boston, MA: National Center for the Study of Adult Learning and Literacy. Viewed 6 February 2006, (http://www.ncsall.net/?id=303).

Crowther, D., Dyson, A. & Millward, A. (2001). Supporting pupils with special educational needs: Issues and dilemmas for special needs coordinators in English primary schools. *European Journal of Special Needs Education, 16, 2,* 85–87.

Cuisenaire, G. & Gattegno, C. (1954). *Numbers in colour: A new method of teaching the processes of arithmetic to all levels of the primary school.* London: Heinemann.

Curriculum Corporation (Australia) (1991). *A national statement on mathematics for Australian schools.* Melbourne: Curriculum Corporation.

Curriculum Corporation (Australia) (1994a). *Mathematics: A curriculum profile for Australian schools.* Melbourne: Curriculum Corporation.

Curriculum Corporation (Australia) (1994b). *Statement on studies of society and environment for Australian schools.* Melbourne: Curriculum Corporation.

Curriculum Council (Western Australia) (1998). *Curriculum framework: Mathematics.* Perth: Curriculum Council. Viewed 6 February 2006, (http://www.curriculum.wa.edu.au/pages/framework/framework08.htm).

Damon, W. & Phelps, E. (1989). Strategic uses of peer learning in children's education. In T.J. Berndt & G.W. Ladd (eds) *Peer relationships in child development* (pp.135–157). New York: Wiley.

Darling-Hammond, L. (2000). Teacher quality and student achievement: A review of state policy evidence. *Educational Policy Analysis Archives, 8,* 1. Viewed 6 February 2006, (http://epaa.asu.edu/epaa/v8n1).

Davies J.D., Garner, P. & Lee, J. (1998). SENCOs and the Code: No longer practising. In J.D. Davies, P. Garner & J. Lee (eds) *Managing special needs in mainstream schools: The role of the SENCO* (pp.36–49). London: Fulton.

Davis, D. & Sorrell, J. (1995). *Mastery learning in public schools.* Valdosta, GA: Valdosta State University. Viewed 6 February 2006, (http://chiron.valdosta.edu/whuitt/files/mastlear.html).

Department of Education and Training (ACT) (1994). *Studies of Society and Environment: Curriculum Framework.* Canberra: Education and Community Services.

Department of Education and Training (ACT) (1997). *Science curriculum framework.* Canberra: DET.

Deschenes, C., Ebeling, D. & Sprague, J. (1999). *Adapting the curriculum in inclusive classrooms.* New York: National Professional Resources.

DEST (Department of Education, Science and Training Australia) (2005). *Teaching reading: National inquiry into the teaching of literacy.* Canberra: Australian Government Publishing Service.

Dettmer, P., Thurston, L. & Dyck, N. (2005). *Consultation, collaboration and teamwork for students with special needs* (5th edn). Boston: Pearson Allyn & Bacon.

Dewey, J. (1933). *How we think: A restatement of the relation of reflective thinking to the educative process.* Boston: Heath.

DfEE (Department for Education and Employment) (1999). *The National Numeracy Strategy: Framework for teaching mathematics.* Sudbury, UK: DfEE.

DfES (Department for Education and Skills) (2001). *Special Educational Needs Code of Practice.* London: DfES. Viewed 8 February 2006, (http://www.teachernet.gov.uk/docbank/index.cfm?id=3724).

DfES (Department for Education and Skills) (2002a). *Framework for teaching science: Years 7, 8 and 9.* London: DfES.

DfES (Department for Education and Skills) (2002b). *Key Stage 3 National Strategy: Developing the science curriculum for pupils with special educational needs.* London: DfES.

DfES (Department for Education and Skills) (2004). *Science module: Induction training for teaching assistants in secondary schools.* London: DfES. Viewed 3 February 2006, (http://www.teachernet.gov.uk/_doc/7133/Sci_Sec.pdf).

DfES (Department for Education and Skills) (2005). *Primary National Strategy: Reviewing the framework for teaching literacy and mathematics.* London: DfES.

DfES (Department for Education and Skills) (2006a). *Key Stage 3 National Strategy: Accessing the National Curriculum for Mathematics.* London: DfES. Viewed 2 February 2006, (http://www.standards.dfes.gov.uk/keystage3/respub/ma_access_nc).

DfES (Department for Education and Skills) (2006b). *Primary national strategy: Framework for teaching mathematics.* London: DfES. Viewed 2 February 2006, (http://www.standards.dfes.gov.uk/primary/publications/mathematics/math_framework).

Dickinson, P. (2003). Whole class interactive teaching. *SET Research for Teachers, 1,* 18–21. Wellington: New Zealand Council for Educational Research.

Dixon, R. & Engelmann, S. (1976). *Spelling through morphographs.* Sydney: SRA/McGraw-Hill.

DO-IT (Disabilities, Opportunities, Internetworking and Technology) (2002). *Working together: Computers and people with learning disabilities.* Seattle: University of Washington.

Dymock, S. & Nicholson, T. (1999). *Reading comprehension: What is it?* Wellington: New Zealand Council for Educational Research.

Dyson, A. & Millward, A. (2000). *Schools and special needs: Issues of innovation and inclusion.* London: Paul Chapman Publishing.

Edwards, C. & Willis, J. (eds) (2005). *Teachers exploring tasks in English language teaching.* Basingstoke, UK: Palgrave Macmillan.

Eggen, P. & Kauchak, D. (2004). *Educational psychology: Windows on classrooms* (6th edn). Upper Saddle River, NJ: Pearson-Merrill.

Elliott, P. & Garnett, C. (1994). Mathematics power for all. In C.A. Thornton & N. Bley (eds) *Windows of opportunity: Mathematics for students with special needs* (pp.3–18). Reston, VA: National Council of Teachers of Mathematics.

Ellis, L.A. (2005). *Balancing approaches: Revisiting the educational psychology research on teaching students with learning difficulties.* Melbourne: Australian Council for Educational Research.

Emanuelsson, I. (2001) Reactive versus proactive support coordinator roles: an international comparison, *European Journal of Special Needs Education, 16,* 2, 133–42.

Engelmann, S. & Bruner, E.C. (1995). *Reading Mastery I: Teacher's Guide.* Columbus, OH: SRA/McGraw-Hill.

ERIC EC (ERIC Clearinghouse on Disabilities and Gifted Education) (1999). *Teaching social studies to students with learning disabilities.* Arlington, VA: Council for Exceptional Children. Viewed 6 February 2006, (http://ericec.org/faq/soc-stud.html).

ERIC EC (ERIC Clearinghouse on Disabilities and Gifted Education) (2003a). *Problem-based learning.* Arlington, VA: Council for Exceptional Children. Viewed 1 February 2006, (http://ericec.org/faq/gt-prob.html)

ERIC EC (ERIC Clearinghouse on Disabilities and Gifted Education) (2003b). *Teaching science to students with disabilities.* USA: Hoagies' Gifted Education Page. Viewed 3 February 2006, (http://www.hoagiesgifted.org/eric/faq/science.html).

Esch, C. (1998). *Project-based and problem-based: The same or different?* Redwood City, CA: Project-Based Learning with Multimedia/San Mateo County Office of Education. Viewed 8 February 2006, (http://pblmm.k12.ca.us/PBLGuide/PBL&PBL.htm).

Esch, C. (2005). *Resource-based learning: Guide to good practice.* Southampton, UK: Subject Centre for Languages, Linguistics and Area Studies. Viewed 1 February 2006, (http://www.lang.ltsn.ac.uk/resources/goodpractice.aspx?resourceid=409).

European Journal of Special Needs Education (2001). 16, 2, June.

Evans. D. (1995). Operant and social learning. In F. Maltby, N. Gage & D. Berliner (eds) *Educational psychology: An Australian and New Zealand perspective* (pp.218–255). Brisbane: Wiley.

Ewing, J. & Miller, D. (2002). A framework for evaluating computer supported collaborative learning. *Educational Technology and Society 5, 1*. Viewed 8 February 2006, (http://ifets.ieee.org/periodical/vol_1_2002/ewing.html).

Farkota, R.M. (2003). Effects of direct instruction on self-efficacy and achievement in mathematics. PhD thesis. Melbourne: Monash University.

Farkota, R.M. (2005). Basic math problems: The brutal reality! *Learning Difficulties Australia Bulletin, 37, 3*, 10–11.

Farrell, P., Balshaw, M. & Polat, F. (2000). The work of the learning support assistants in mainstream schools: Implications for educational psychologists, *Educational and Child Psychology, 17, 2*, 66–76.

Fenton, A. (2002). Supporting inclusive science for special educational needs. *Education in Science, 196*, 8–9.

Fisher, E. (2002). Science in a special school setting: Strategies from Charlton School. *Support for Learning, 17, 4*, 162–167.

Flavell, J.S. & Wakelam, B.B. (1960). *Primary mathematics: Introduction to the language of number.* London: Methuen.

Fletcher-Campbell, F. & Cullen, M.A. (2000). Schools' perceptions of support services for special educational needs. *Support for Learning, 15, 2*, 90–94.

Foreman, P., Bourke, S. & Mishra, G. (2001). Assessing the support needs of children with disability in regular classes. *International Journal of Disability and Education, 48, 3*, 239–252.

Forlin, C. (2000). *A report on the role of the Support Teacher (Learning Difficulties) in regular schools in Queensland.* Occasional Paper. Toowoomba, Qld: Centre for Educational Research and Development, University of Southern Queensland.

Forlin, C. (2001a). Inclusion: Identifying potential stressors for regular class teachers. *Educational Research, 43, 3*, 235–245.

Forlin, C. (2001b). The role of the support teacher in Australia. *European Journal of Special Needs Education, 16, 2*, 121–131.

Foster, C. (2004). Anchored instruction. In B. Hoffman (ed.) *Encyclopedia of Educational Technology.* San Diego: San Diego State University. Viewed 1 February 2006, (http://coe.sdsu.edu/eet/articles/anchoredinstruc/start.htm).

Fox, G. (2003). *A handbook for learning support assistants* (2nd edn). London: Fulton.

Freiberg, H.J. & Driscoll, A. (2005). *Universal teaching strategies* (4th edn). Boston: Allyn & Bacon.

Frostig, M. & Horne, D. (1964). *The Frostig program for development of visual perception.* Chicago: Follett.

Fuchs, D., Fuchs, L.S. & Burish, P. (2000). Peer-assisted learning strategies: An evidence-based practice to promote reading achievement. *Learning Disabilities Research and Practice, 15, 2*, 85–91.

Galton, M., Hargreaves, L., Comber, W., Wall, D. & Pell, A. (1999). *Inside the primary classroom, 20 years on.* London: Routledge.

Garner, S. (2005). The use of task based learning in a Web commerce development undergraduate unit. In *Proceedings of World Conference on Educational Multimedia, Hypermedia and Telecommunications 2005* (pp. 3117–3122). Norfolk, VA: Association for the Advancement of Computing in Education.

Gattegno, C. (1960). *Modern mathematics with numbers in colour.* Reading, UK: Cuisenaire Company.

Gaustad, J. (1993). Peer and cross-age tutoring. *ERIC Digest 79*. Eugene, OR: Clearinghouse on Educational Policy and Management. Viewed 8 February 2006, (http://cepm.uoregon.edu/publications/digests/digest079.html).

Gentile, J.R. & Lalley, J.P. (2003). *Standards and mastery learning.* Thousand Oaks, CA: Corwin.

Giangreco, M., Edelman, S. & Broer, S. (2001). Respect, appreciation, and acknowledgement of paraprofessionals who support students with disabilities. *Exceptional Children, 67, 4*, 485–498.

Good, T.L. & Brophy, J.E. (1997). *Looking in classrooms* (7th edn). New York: Longman.

Good, T.L. & Brophy, J.E. (2003). *Looking in classrooms* (9th edn). Boston: Allyn & Bacon.

Graesser, A. & Person, N. (1994). Question asking during tutoring. *American Educational Research Journal, 31*, 104–137.

Graham, L. & Bellert, A. (2005). Reading comprehension difficulties experienced by students with learning disabilities. *Australian Journal of Learning Disabilities, 10, 2*, 71–78.

Graham, S. (2000). Should the natural learning approach replace spelling instruction? *Journal of Educational Psychology, 92, 2*, 235–247.

Graham, S. & Harris, K.R. (2000). The role of self-regulation and transcription skills in writing and writing development. *Educational Psychologist, 35*, 3–12.

Graham, S., Harris, K., MacArthur, C. & Schwartz, S. (1991). Writing and writing instruction for students with learning disabilities: Review of a research program. *Learning Disabilities Quarterly, 14*, 89–114.

Graves, D.H. (1983). *Writing: Teachers and children at work.* Exeter, NH: Heinemann.

Greenwood, C.R. (1991). Classwide peer tutoring: Longitudinal effects on the reading, language, and mathematics achievement of at-risk students. *Reading, Writing and Learning Disabilities, 7, 2*, 105–123.

Greenwood, C.R. & Delquadri, J. (1995). Classwide peer tutoring and the prevention of school failure. *Preventing School Failure, 39, 4*, 21–25.

Greenwood, C.R., Delquadri, J.C. & Hall, R.V. (1989). Longitudinal effects of classwide peer tutoring. *Journal of Educational Psychology, 81, 3*, 371–383.

Gregg, N. & Mather, N. (2002). School is fun at recess: Informal analyses of written language for students with learning disabilities. *Journal of Learning Disabilities 35, 1*, 7–22.

Gross, P.R. (2005). *Less than proficient: A review of the draft science framework for the 2009 National Assessment of Educational Progress.* Washington, DC: Thomas B. Fordham Foundation.

Gummett, B. & Martin, C. (2001). Geography. In B. Carpenter, R. Ashdown & K. Bovair (eds) *Enabling access: Effective teaching and learning for pupils with learning difficulties* (2nd edn, pp.90–99). London: Fulton.

Hackling, M.W., Goodrum, D. & Rennie, L.J. (2001). The state of science in Australian secondary schools. *Australian Science Teachers' Journal, 47, 4*, 6–18.

Hall, S. (1997). The problem with differentiation. *School Science Review, 78, 284*, 95–98.

Hammann, L. & Stevens, R. (2003). Instructional approaches to improving students' writing of compare-contrast essays: An experimental study. *Journal of Literacy Research, 35, 2*, 731–756.

Hanrahan, M.U., Cooper, T.J. & Russell, A.L. (1997). *Science for all: Action researching literacy difficulties in a Year 8 science class.* Paper presented to the World Congress of Participatory Convergence in Knowledge, Space and Time, Cartagena, Colombia, June.

Hardman, F., Smith, F. & Wall, K. (2003). Interactive whole-class teaching in the National Literacy Strategy. *Cambridge Journal of Education, 33, 2*, 197–215.

Hargreaves, L., Moyles, J., Merry, R., Paterson, F., Pell, A. & Esartes-Sarries, V. (2003). How do primary school teachers define and implement interactive teaching in the National Literacy Strategy in England? *Research Papers in Education, 18, 3*, 217–236.

Harniss, M.K., Dickson, S.V., Kinder, D. & Hollenbeck, K.L. (2001). Textual problems and instructional solutions: Strategies for enhancing learning from published history textbooks. *Reading and Writing Quarterly, 17*, 127–150.

Harniss, M.K., Carnine, D.W., Silbert, J. & Dixon, R.C. (2002). Effective strategies for teaching mathematics. In E.J. Kameenui & D.C. Simmons (eds) *Effective teaching strategies that accommodate diverse learners* (2nd edn, pp.121–148). Upper Saddle River, NJ: Merrill-Prentice Hall.

Harrington, A. (1999). *Whole class interactive teaching in mathematics.* London: Training & Development Agency for Schools. Viewed 8 February 2006, (www.tda.gov.uk/upload/resources/doc/a/aidan-harrington.doc).

Harrington, M.O., Paisey, T.J., Israel, M.L., Langford, E.G. & Holland, E.E. (2004). *Improved standard scores in spelling utilizing precision teaching with computer self-instruction software and reward systems.* Canton, MA: Judge Rotenberg Educational Center. Viewed 8 February 2006, (http://www.effectivetreatment.org/improved_standard.html).

Harris, A. (1998). Effective teaching: A review of the literature. *School Leadership and Management, 18, 2,* 169–183.

Harris, K., Graham, S. & Mason, L.H. (2003). Self-regulated strategy development in the classroom: Part of a balanced approach to writing instruction for students with disabilities. *Focus on Exceptional Children, 35, 7,* 1–16.

Harris, R.J. & Luff, I. (2004). *Meeting special needs in history.* London: Fulton.

Hattie, J.A. (2003). *Teachers make a difference: What is the research evidence?* Background paper to an address to the ACER Research Conference, Melbourne, October 19–23.

Healey, M., Roberts, C., Jenkins, A. & Leach, J. (2002). Disabled students and fieldwork: Towards inclusivity? *Planet, 3,* 9–10.

Heeks, P. & Kinnell, M. (1997). *Learning support for special educational needs.* London: Taylor Graham.

Heimlich, J.E. & Daudi, S.S. (1996). Environmental education and learners with disabilities. *EETAP Resource Library, 1.* April issue. Viewed 3 February 2006, (http://eelink.net/eetap/info1.pdf).

Heinz, S.A. (2000). *The power of the multisensory approach.* Bellevue, WA: Slingerland Institute for Literacy.

Hemmens, A. (1999). Learning through the senses. *Primary Science Review, 59,* 20–23.

Hempenstall, K. (1996). The gulf between educational research and policy: The example of direct instruction and whole language. *Behaviour Change, 13, 1,* 33–46.

Henderson, J. & Wellington, J. (1998). Lowering the language barrier in learning and teaching science. *School Science Review, 79, 288,* 35–46.

Henry, M.K. (1998). Structured, sequential, multisensory teaching: The Orton legacy. *Annals of Dyslexia, 48,* 3–27.

Hess, M. & Wheldall, K. (1999). Strategies for improving the written expression of primary children with poor writing skills. *Australian Journal of Learning Disabilities, 4, 4,* 14–20.

Higher Education Academy Subject Centre for Geography, Earth and Environmental Sciences (2006). Plymouth, UK: Higher Education Academy. Viewed 3 February 2006, (http://www.gees.ac.uk).

HMI (Her Majesty's Inspector of Schools) (1997). *The SEN Code of Practice: Two years on.* London: OfSTED.

Hoffman, J.V., Baumann, J. & Afflerbach, P. (2000). *Balancing principles for teaching elementary reading.* Mahwah, NJ: Erlbaum.

Hopkins, S. (1998). Learning disabled (LD) performance under pressure. In D. Greaves & P. Jeffery (eds) *Strategies for intervention with special needs students* (pp.43–61). Melbourne: Australian Resource Educators' Association.

Houghton Mifflin (n.d.). *Project based learning.* Wilmington, MA: Houghton Mifflin Online Study Center. Viewed 1 February 2006, (http://college.hmco.com/education/resources/res_project/students/background.html).

House of Commons Education and Skills Committee (2005). *Teaching children to read.* London: HMSO.

Howell, K.W. (1995). Learning styles instruction: Questions and answers about aptitude by treatment interactions. *Special Education Perspectives, 4, 1,* 11–15.

Hudson, P. (1996). Using a learning set to increase the test performance of students with learning disabilities in social studies classes. *Learning Disabilities Research and Practice, 11, 2,* 78–85.

Hudson, P. (1997). Using teacher-guided practice to help students with learning disabilities acquire and retain social studies content. *Learning Disability Quarterly, 20, 1,* 23–32.

Huitt, W. (1996). *Mastery learning.* Educational Psychology Interactive. Valdosta, GA: Valdosta State University. Viewed 8 February 2006, (http://chiron.valdosta.edu/whuitt/col/instruct/mastery.html).

Hummel, S. (2000). Developing comprehension skills of secondary students with specific learning difficulties. *Australian Journal of Learning Disabilities, 5, 4,* 22–27.

Hunter, R. (ed.) (2004). *Madeline Hunter's Mastery Teaching: Increasing instructional effectiveness in elementary and secondary schools.* Thousand Oaks, CA: Corwin.

Hunting, R.P. (1996). Does it matter if Mary can read but can't add up? *Education Australia, 33,*16–19.

IDA (International Dyslexia Association) (2000). *Multisensory teaching.* San Francisco: IDA. Viewed 8 February 2006, (http://www.dyslexia-ncbida.org/articles/feb05/multisensory_teaching.html).

Inclusive Science and Special Education Needs (n.d.). Hatfield, UK: Association for Science Education. Viewed 3 February 2006, (http://www.issen.org.uk).

Inclusive science and special educational needs resources (2003). CD-ROM. Hatfield, UK: Association for Science Education.

Isaacs, N. (1960). *New light on children's ideas of number: The work of Professor Piaget.* London: Educational Supply Association.

Itard, J.M.G. (1962). *The wild boy of Aveyron* (1801/1806). Trans. G. Humphrey & M. Humphrey. New York: Appleton-Century-Crofts.

Jacobs, J. (2004). The limits of 'discovery learning'. Message board, 4 February, joannejacobs.com. Viewed 1 February 2006, (http://www.joannejacobs.com/mtarchives/013751.html).

Jacobsen, D.A., Eggen, P. & Kauchak, D. (2002). *Methods for teaching* (6th edn). Upper Saddle River, NJ: Merrill-Prentice Hall.

Janney, R. & Snell, M.E. (2004). *Modifying schoolwork* (2nd edn). Baltimore: Brookes.

Jitendra, A.K., Nolet, V., Xin, Y.P. & Da Costa, J. (2001). An analysis of middle school geography textbooks: Implications for students with learning problems. *Reading and Writing Quarterly, 17,* 151–173.

Johnson, D.W. & Johnson, R.T. (1999). *Learning together and alone* (5th edn). Edina, MN: Interaction Book Company.

Johnson, D.W., Johnson, R.T. & Holubec, E.J. (2002). *Circles of learning* (5th edn). Edina, MN: Interaction Book Company.

Johnson, D.W., Johnson, R.T. & Stanne, M.B. (2000). *Cooperative learning methods: A meta-analysis.* Minneapolis: Cooperative Learning Center, University of Minnesota. Viewed 8 February 2006, (http://www.co-operation.org/pages/cl-methods.html).

Johnston, R. & Watson, J. (2005). *The effects of synthetic phonics teaching on reading and spelling attainment: A seven year longitudinal study.* Edinburgh: Scottish Executive Education Department. Viewed 8 February 2006, (http://www.scotland.gov.uk/library5/education/sptrs.pdf).

Jones, E.D. & Southern, W.T. (2003). Balancing the perspectives on mathematics instruction. *Focus on Exceptional Children, 35, 9,* 1–16.

Jones, R.C. (2001). *Strategies for reading comprehension: Reciprocal teaching.* Winston-Salem, NC: ReadingQuest, Wake Forest University. Viewed 1 February 2006, (http://curry.edschool.virginia.edu/go/readquest/strat/rt.html).

Jones, S. & Tanner, H. (2005). *Teachers' interpretations of effective whole class interactive teaching.* Palermo, Italy: Gruppo di Ricerca sull'Insegnamento delle Matematiche, Università degli Studi di Palermo. Viewed 8 February 2006, (http://math.unipa.it/~grim/ATanner122-127.PDF).

Joyce, B., Weil, M. & Calhoun, E. (2004). *Models of teaching* (7th edn). Boston: Allyn & Bacon.

Kalkowski, P. (2001). Peer and cross-age tutoring. Close *Up #18: School Improvement Research Series.* Portland, OR: NW Regional Educational Laboratory.

Kauchak, D. and Eggen, P.D. (2003). *Learning and teaching: Research-based methods* (4th edn). Boston: Allyn & Bacon.

Kavale, K. & Forness, S. (2000). Policy decisions in special education: The role of meta-analysis. In R. Gersten, E. Schiller & S. Vaughn (eds). *Contemporary special education research* (pp.281–326). Mahwah, NJ: Erlbaum.

Kennedy, G. (2000). Students with special needs and the Web. *Australian Journal of Learning Disabilities, 5, 3,* 39.

Kershner, R. (2000). Teaching children whose progress in learning is causing concern. In D. Whitebread (ed.) *The psychology of teaching and learning in the primary school* (pp.277–299). London: Routledge-Falmer.

Killen, R. (1998). *Effective teaching strategies: Lessons from research and practice* (2nd edn). Wentworth Falls, NSW: Social Science Press.

Kilpatrick, J., Swafford, J. & Findell, B. (eds) (2001). *Adding it up: Helping children learn mathematics.* Washington, DC: National Academy Press.

Kindsvatter, R., Wilen, W. & Ishler, M. (1996). *Dynamics of effective teaching* (3rd edn). New York: Longman.

King, H. (2001). Case studies in problem-based learning from geography, earth and environmental sciences. *Planet, Special Edition 2,* 3–4.

Kingham, P.H. & Blackmore, A.M. (2003). Computer based instruction in blending is effective for Year 2 children with reading difficulties. *Australian Journal of Learning Disabilities, 8, 1,* 30–37.

Kirk, S. & Kirk, W. (1971). *Psycholinguistic learning disabilities: Diagnosis and remediation.* Urbana: University of Illinois Press.

Kizlik, B. (2005). *System for instruction — ADPRIMA.* Florida: ADPRIMA. Viewed 8 February 2006, (http://www.adprima.com/adprisys.htm).

Klein, D., Braams, B.J., Parker, T., Quirk, W., Schmid, W. & Wilson, W.S. (2005). *The state of State math standards: 2005.* Washington, DC: Thomas B. Fordham Foundation.

Kohlberg, L. (1978). Revisions in the theory and practice of moral development. In W. Damon (ed.) *Moral development.* San Francisco: Jossey-Bass.

Kozloff, M. (2003) *Main features of Direct Instruction.* Wilmington, NC: University of North Carolina. Viewed 8 February 2006, (http://people.uncw.edu/kozloffm/difeatures.doc).

Kraft, N. (2005). *Criteria for authentic project-based learning.* Denver, CO: RMC Research Corporation. Viewed 8 February 2006, (http://www.rmcdenver.com/useguide/pbl.htm).

Kubina, R.M. & Starlin, C.M. (2003). Reading with precision. *European Journal of Behaviour Analysis, 4, 1/2,* 13–21.

Lacey, P. (2001). *Support partnerships: Collaboration in action.* London: Fulton.

Lam, B.F.Y. & Westwood, P. (2006). Spelling and ESL learners: A strategy-training approach. *Special Education Perspectives 15, 1,* 11–23.

Lamb, J. (2004). The impact of teachers' understanding of division on their students' knowledge of division. In B.A. Knight & W. Scott (eds) *Learning difficulties: Multiple perspectives* (pp.153–169). Frenchs Forest, NSW: Pearson.

Lancaster, J. (2005). Is it really possible? Can students with learning difficulties ever achieve higher levels of self-efficacy? *Special Education Perspectives, 14, 2,* 46–61.

Lave, J. and Wenger, E. (1991). *Situated learning: Legitimate peripheral participation.* Cambridge: University of Cambridge Press.

Learning and Teaching Scotland (2005). *Synthetic v. analytic phonics.* Dundee: Learning and Teaching Scotland. Viewed 2 February 2006, (http://www.ltscotland. org.uk/5to14/specialfocus/earlyintervention/issues/phonics.asp).

Lee, C. (2001). Problem-based learning: A personal view. *Planet, Special Edition 2,* 10.

Leiding, D. (2002). *The 'won't' learners: An answer to their cry.* Lanham, MD: Scarecrow Press.

Lembo, J.M. (1971). *Why teachers fail.* Columbus: Merrill.

Leming, J.S. (2003). Ignorant activists: Social change, 'higher order thinking', and the failure of social studies. In J. Leming, L. Ellington & K. Porter-Magee (eds) *Where did social studies go wrong?* (pp.124–142). Washington, DC: Thomas B. Fordham Foundation.

Lenz, K. & Schumaker, J. (1999). *Adapting language arts, social studies, and science materials for the inclusive classroom.* Reston, VA: Council for Exceptional Children.

Lerner, J. & Kline, F. (2006). *Learning disabilities and related disorders: Characteristics and teaching strategies* (10th edn). Boston: Houghton Mifflin.

Linden, L., Banerjee, A. & Duflo, E. (2003). Computer-assisted learning: Evidence from a randomized experiment. Cambridge, MA: Poverty Action Lab, Massachusetts Institute of Technology. Viewed 8 February, (http://www.povertyactionlab.com/ papers/banerjee_duflo_linden.pdf).

Lindsley, O.R. (1964). Direct measurement and prosthesis of retarded behavior. *Journal of Education, 147,* 62–81.

Lindsley, O.R. (1990). Precision teaching. *Teaching Exceptional Children, 22, 3,* 10–15.

Lindsley, O.R. (1992a). Precision teaching: Discoveries and effects. *Journal of Applied Behavior Analysis, 25, 1,* 51–57.

Lindsley, O.R. (1992b). Why aren't effective teaching tools widely adopted? *Journal of Applied Behavior Analysis, 25, 1,* 21–26.

Lo, M.L., Morris, P. & Che, M.W. (2000). Catering for diversity. In B. Adamson, T.K.L. Kwan & K.K. Chan (eds) *Changing the curriculum: The impact of reform on primary schooling in Hong Kong* (pp.217–242). Hong Kong: University of Hong Kong Press.

Lockery, M. & Maggs, A. (1982). Direct instruction research in Australia: A ten-year analysis. *Educational Psychology, 2, 3,* 263–288.

Lovell, K. (1978). *The growth of basic mathematical and scientific concepts in children* (6th edn). London: University of London Press.

Lovell, K. & Smith, C.H.J. (1956). *Two-Grade Arithmetic.* Aylesbury: Odhams.

Lovitt, T.C. & Horton, S.V. (1994). Strategies for adapting science textbooks for youth with learning disabilities. *Remedial and Special Education, 15, 2,* 105–116.

Lubliner, S. & Smetana, L. (2005). Effects of comprehensive vocabulary instruction on Title I students' metacognitive word-learning skills and reading comprehension. *Journal of Literacy Research, 37, 2,* 163–200.

McBer, H. (2000). *Research into teacher effectiveness.* London: Department for Education and Employment.

McCarthy, C.B. (2005). Effects of thematic-based, hands-on science teaching versus a textbook approach for students with disabilities. *Journal of Research in Science Teaching, 42, 3,* 245–263.

McCleery, J. A. & Tindal, G.A. (1999). Teaching the scientific method to at-risk students and students with learning disabilities through concept anchoring and explicit instruction. *Remedial and Special Education, 20, 1,* 7–18.

McInerney, D.M. & McInerney, V. (2002). *Educational psychology: Constructing learning* (3rd edn). Sydney: Prentice Hall.

McIntosh, A. & Dole, S. (2000). Number sense and mental computation: Implications for numeracy. In *ACER Research Conference 2000: Improving numeracy learning* (pp.34–37). Melbourne: Australian Council for Educational Research.

McKinnon, M. & Rigby, N. (2004). *Task-based learning.* Basingstoke, UK: Onestopenglish/Macmillan Publishers. Viewed 1 February 2006, (http://www.onestopenglish.com/teacher_support/methodology/archive/teaching-approaches/task_based_learning.htm)

McLaren, P. (2003). *Life in schools: An introduction to critical pedagogy in the foundations of education* (4th edn). Boston: Allyn & Bacon

McLellan, H. (1995). *Situated learning perspectives.* Englewood Cliffs, NJ: Educational Technology Publications.

McNaughton, D., Hughes, C. & Clark, K. (1994). Spelling instruction for students with learning disabilities: Implications for research and practice. *Learning Disability Quarterly, 17, 3,* 169–185.

MacIver, M.A. & Kemper, E. (2002). Research on Direct Instruction in reading. *Journal of Education for Students Placed at Risk, 7, 2,* 107–116.

Mackintosh, M. (2005). *GTIP Think Piece: Geography and numeracy.* Sheffield, UK: Geographical Association. Viewed 8 February 2006, (http://www.geography.org.uk/projects/gtip/thinkpieces/numeracy)

Maheady, L., Harper, G.F. & Mallette, B. (2001). Peer-mediated instruction and interventions and students with mild disabilities. *Remedial and Special Education, 22, 1,* 4–14.

Marston, D. (1996). A comparison of inclusion only, pull-out only, and combined service models for students with mild disabilities, *Journal of Special Education, 30, 2,* 121–32.

Martin, A.J. & Marsh, H.W. (2003). Fear of failure: Friend or foe? *Australian Psychologist, 38, 1,* 31–38.

Martin, C. & Gummett, B. (2001). History. In B. Carpenter, R. Ashdown & K. Bovair (eds) *Enabling access: Effective teaching and learning for pupils with learning difficulties* (2nd edn, pp.80–89). London: Fulton.

Marvin, M.L. & Stokoe, C. (2003). *Access to science: Curriculum planning and practical activities for pupils with learning difficulties.* London: Fulton.

Mastropieri, M.A. & Scruggs, T.E. (1993). *A practical guide for teaching science to students with special needs in inclusive settings.* Austin, TX: ProEd.

Mastropieri, M.A., Scruggs, T.E. & Butcher, K. (1997). How effective is inquiry learning for students with mild disabilities? *Journal of Special Education, 31, 2,* 199–211.

Micallef, S. & Prior, M. (2004). Arithmetic learning difficulties in children. *Educational Psychology, 24, 2,* 175–200.

Miller, S.P. (1999). Teaching initial math skills to students with learning disabilities. In D. Barwood, D. Greaves & P. Jeffery (eds). *Teaching numeracy and literacy: Interventions and strategies for 'at risk' students* (pp.165–173). Melbourne: Australian Resource Educators' Association.

Minton, P. (2002). Using information and communication technology to help dyslexics and others learn to spell. *Australian Journal of Learning Disabilities, 7, 3,* 26–31.

Montague, M. & Bos, C. S. (1986). The effects of cognitive strategy training on verbal math problem solving. *Journal of Learning Disabilities, 19, 1,* 26–33.

Montessori, M. (1919). *The advanced Montessori method: Scientific pedagogy as applied to the education of children from seven to eleven years.* London: Heinemann.

Morcom, N. (2005). Effective reading instruction. *Special Education Perspectives, 14, 2,* 5–17.

Morgan, C. & Morris, G. (1999). *Good teaching and learning,* Buckingham, UK: Open University Press.

Morgan, J. & Ashbaker, B.Y. (2001). *A teacher's guide to working with paraeducators and other classroom aides.* Alexandria, VA: Association for Supervision and Curriculum Development.

National Council for the Social Studies (1994). *Expectations of excellence: Curriculum standards for social studies.* Silver Springs, MD: NCSS.

National Council for the Social Studies (2005). Silver Springs, MD: NCSS. Viewed 8 February 2006, (http://www.socialstudies.org).

National Reading Panel (2000). *Teaching children to read: An evidence-based assessment of the scientific research literature on reading and its implications for reading instruction.* Washington, DC: National Institute of Child Health and Human Development. Viewed 8 February 2006, (http://www.nichd.nih.gov/publications/nrp/smallbook. pdf).

National Science Teachers' Association (NSTA) (2004). *NSTA Position statement: Students with disabilities.* Washington, DC: The National Science Teachers' Association. Viewed 8 February 2006, (http://www.nsta.org/disabilities).

NCC (National Curriculum Council) (1992). *Teaching science to pupils with special educational needs.* York, UK: NCC.

NCTM (National Council of Teachers of Mathematics) (2005a). Reston, VA: NCTM. Viewed 2 February 2006, (http://www.nctm.org).

NCTM (National Council of Teachers of Mathematics (US) (2005b). *Number and Operations Standard.* Reston, VA: NCTM. Viewed 2 February 2006, (http://standards.nctm.org/document/appendix/numb.htm).

NCTM (National Council of Teachers of Mathematics) (2005c). *Overview: Principles for school mathematics.* Reston, VA: NCTM. Viewed 8 February 2006, (http://standards.nctm.org/document/chapter2/index.htm).

Neill, M. (2005). Total recall. *Special Children, 164,* 22–25.

New Horizons for Learning (2005), *Cooperative learning.* Seattle: New Horizons for Learning. Viewed 1 February 2006, (http://www.newhorizons.org/strategies/cooperative/front_cooperative.htm).

Newman, F. (1990). Qualities of thoughtful social studies classes: An empirical profile. *Journal of Curriculum Studies, 22,* 253–275.

Newman, F. (1991). Higher-order thinking in the teaching of social studies. In J.F. Voss, D.N. Perkins & J.W. Segal (eds) *Informal reasoning and education* (pp.381–400). Hillsdale, NJ: Erlbaum.

Norman, K.L. (1997). *Teaching in the switched on classroom: An introduction to electronic education and HyperCourseware.* College Park, MD: University of Maryland. Viewed 31 January 2006, (http://www.lap.umd.edu/SOC/sochome.html).

NYSED (New York State Education Department) (2005). *Social studies resource guide with core curriculum.* Albany, NY: State Education Department, University of the State of New York. Viewed 8 February 2006, (http://www.emsc.nysed.gov/ciai/socst/ssrg.html).

OECD (Organization for Economic Co-operation & Development) (2000). *Measuring students' knowledge and skills: The PISA 2000 assessment of reading, mathematical and scientific literacy.* Paris: OECD. Viewed 8 February 2006, (http://www.pisa.oecd.org/dataoecd/44/63/33692793.pdf).

OfSTED (Office for Standards in Education) (2004). *Outdoor education: Aspects of good practice (HMI ref. 2151).* London: Her Majesty's Inspectors.

OfSTED (Office for Standards in Education) (2005a). *Art and design, English, history and music top of the class in secondary schools.* Media release. London: OfSTED, 2 February.

OfSTED (Office for Standards in Education) (2005b). *Subject reports 2003/04: Geography in primary schools (HMI ref. 2344).* London: Her Majesty's Inspectors.

Okolo, C.M. & Ferretti, R.P. (1996). Knowledge acquisition and technology-supported projects in the social studies for students with learning disabilities. *Journal of Special Education Technology, 13, 2*, 91–103.

Ormrod, J.E. (2000). *Educational psychology: Developing learners* (3rd edn). Upper Saddle River, NJ: Merrill-Prentice Hall.

OTEC (Oregon Technology in Education Council) (2005). *Learning theories and transfer of learning.* Eugene, OR: OTEC. Viewed 8 February 2006, (http://otec.uoregon.edu/learning_theory.htm).

Owen, R.L. & Fuchs, L.S. (2002). Mathematical problem-solving strategy instruction for third-grade students with learning disabilities. *Remedial and Special Education, 23, 5*, 268–278.

Palincsar, A. & Brown, A.L. (1984). Reciprocal teaching of comprehension-fostering and comprehension-monitoring activities. *Cognition and Instruction, 1*, 117–175.

Parmar, R.S., Cawley, J.F. & Frazita, R.R. (1996). Word-problem solving by students with and without mild disabilities, *Exceptional Children 62, 5*, 415–429.

Peters, M.L (1985). *Spelling: Caught or taught? A new look.* London: Routledge.

Phillips, D. (2004). Writing with a word processor. *SET: Research Information for Teachers, 3*, 6–8.

Piaget, J. (1963). *Origins of intelligence in children.* New York: Norton.

Piaget, J. (1965). *The moral judgment of the child.* New York: Free Press.

Piaget, J. (1971). *The child's conception of the world* (1929). London: Routledge & Kegan Paul.

Pikulski, J.J. & Chard, D.J. (2005). Fluency: Bridge between decoding and reading comprehension. *The Reading Teacher, 58, 6*, 510–519.

Pincott, R. (2004). Are we responsible for our children's maths difficulties? In B.A. Knight & W. Scott (eds) *Learning difficulties: Multiple perspectives* (pp.129–140). Frenchs Forest, NSW: Pearson.

Planet (2001). Special Edition 2, November. Viewed 1 February 2006, (http://www.gees.ac.uk/pubs/planet/pbl.pdf).

Polkinghorne, J. (2004). Electronic literacy: Part 1 & Part 2. *Australian Journal of Learning Disabilities, 9, 2*, 24–27.

Pollington, M.F., Wilcox, B. & Morrison, T.G. (2001). Self-perception in writing: The effects of writing workshop and traditional instruction on intermediate grade students. *Reading Psychology 22*, 249–265.

Polloway, E.A. & Patton, J.R. (1997). *Strategies for teaching learners with special needs* (6th edn). Upper Saddle River, NJ: Merrill-Prentice Hall.

Polloway, E.A., Patton, R.J & Serna, L. (2005). *Strategies for teaching learners with special needs* (8th edn). Upper Saddle River, NJ: Merrill-Prentice Hall.

Pressley, M. (1999). Self-regulated comprehension processing and its development through instruction. In L. Gambrell, L.M. Morrow, S.B. Neuman & M. Pressley (eds) *Best practices in literacy instruction* (pp.90–97). New York: Guilford Press.

Pressley, M. & McCormick, C.B. (1995). *Advanced educational psychology for educators, researchers and policymakers.* New York: HarperCollins.

Pressley, M. & Woloshyn, V. (1995). *Cognitive strategy instruction that really improves children's academic performance* (2nd edn). Cambridge, MA: Brookline Books.

Prideaux, A., Marsh, K.A. & Caplygin, D. (2005). Efficacy of the Cellfield Intervention for reading difficulties: An integrated computer-based approach targeting deficits associated with dyslexia. *Australian Journal of Learning Disabilities, 10, 2*, 51–62.

Project PRODUCT (n.d.). *Blueprint for PRODUCTive classrooms: A guide for using chart-based learning to structure your teaching.* Shawnee Mission, KS: Project PRODUCT, Kansas State Department of Education. Viewed 30 January 2006, (http://www.celeration.org/pdf/blueprint.pdf).

Purdie, N. & Ellis, L. (2005). *A review of the empirical evidence identifying effective interventions and teaching practices for students with learning difficulties in Years 4, 5 and 6.* Melbourne: Australian Council for Educational Research.

QCA (Qualifications and Curriculum Authority) (2001a). *Mathematics: Introduction.* London: QCA. Viewed 2 February 2006, (http://www.ncaction.org.uk/subjects/maths/index.htm).

QCA (Qualifications and Curriculum Authority) (2001b). *Mathematics: The level descriptions.* London: QCA. Viewed 2 February 2006, (http://www.ncaction.org.uk/subjects/maths/levels.htm).

QCA (Qualifications and Curriculum Authority) (2001c). *Planning, teaching and assessing the curriculum for pupils with learning difficulties: General Guidelines.* London: QCA. Viewed 3 February 2006, (http://www.nc.uk.net/ld/GG_content.html).

QCA (Qualifications and Curriculum Authority) (2001d). *Planning, teaching and assessing the curriculum for pupils with learning difficulties: Geography.* London: QCA. Viewed 3 February 2006, (http://www.nc.uk.net/ld/Ge_content.html)

QCA (Qualifications and Curriculum Authority) (2001e). *Planning, teaching and assessing the curriculum for pupils with learning difficulties: History.* London: QCA. Viewed 3 February 2006, (http://www.nc.uk.net/ld/Hi_content.html).

QCA (Qualifications and Curriculum Authority) (2001f). *Planning, teaching and assessing the curriculum for pupils with learning difficulties: Science.* London: QCA. Viewed 3 February 2006, (http://www.nc.uk.net/ld/Sc_content.html).

QCA (Qualifications and Curriculum Authority) (2005). *The importance of mathematics.* London: QCA. Viewed 3 February, (http://www.nc.uk.net/nc/contents/Ma-home.htm).

Queensland Department of Education and the Arts (1998). *Studies of Society and Environment.* Brisbane: Queensland Department of Education and the Arts. Viewed 8 February 2006, (http://education.qld.gov.au/curriculum/area/sose).

Rakes, T.A., Choate, J. & Stringer, G.L. (2004). Essential science: Relevant topics, process, and strategies. In J. Choate (ed.) *Successful inclusive teaching: Proven ways to detect and correct special needs* (4th edn, pp.312–343). Boston: Allyn & Bacon.

Regina Public Schools and Saskatchewan Learning (2003). *Inquiry learning.* Regina, Canada: Saskatchewan Learning. Viewed 8 February 2006, (http://wblrd.sk.ca/~bestpractice/inquiry/index.html).

Reynolds, D. & Farrell, S. (1996). *Worlds apart? A review of international studies of educational achievement involving England.* London: HMSO.

Riley, J. & Reedy, D. (2000). *Developing writing for different purposes.* London: Paul Chapman.

Ritchie, R. (2001). Science. In B. Carpenter, R. Ashdown & K. Bovair (eds) *Enabling access: Effective teaching and learning for pupils with learning difficulties* (2nd edn, pp.52–65). London: Fulton.

Roach, P. (2005). Does precision monitoring work? *Living Psychology, 3,* Autumn issue. Viewed 8 February, (http://www.buckscc.gov.uk/schools/childrens_services/educational_psychology_service/vol_3/does_precision_monitoring_work.asp).

Robertson, P., Hamill, P. & Hewitt, C. (1994). Interchange 2: Effective support for learning: Themes from the RAISE Project. Edinburgh: Research and Intelligence Unit, Department of Education.

Rose, J. (2005). *Independent review of the teaching of early reading: Interim report.* London: Department for Education and Skills.

Rosenberg, M.S., O'Shea, L. & O'Shea, D. (2006). *Student teacher to master teacher: A practical guide for educating students with special needs.* Upper Saddle River, NJ: Pearson-Merrill-Prentice Hall.

Rosenshine, B. & Meister, C. (1994). Reciprocal teaching: A review of research. *Review of Educational Research 64, 4,* 479–530.

Rosenshine, B. & Meister, C. (1995). Direct instruction. In L.W. Anderson (ed.) *International encyclopedia of teaching and teacher education* (2nd edn, pp.143–149). New York: Pergamon.

Rosenshine, B. & Stevens, R. (1986). Teaching functions. In M.C. Wittrock (ed.) *Handbook of research on teaching* (3rd edn, pp.376–391). New York: Macmillan.

Rousseau, J.J. (1979). *Emile: or, on education* (1762). Trans. A. Bloom. New York: Basic Books.

Rowe, K.R. (2004). The importance of teaching: Ensuring better schooling by building teacher capacities that maximize the quality of teaching and learning provision — implications of findings from the international and Australian evidence-based research. Paper presented to the Making Schools Better Conference, Melbourne, 26–27 August 2004. Viewed 11 February 2006, (http://www.acer.edu.au/research/programs/documents/RowePaperMakingSchoolsBetter26_27_Aug2004.pdf).

Rowe, M.B. (1978). *Teaching science as continuous enquiry.* New York: McGraw-Hill.

Rowe, M.B. (1986). Wait time: slowing down may be a way of speeding up. *Journal of Teacher Education, 37, 1,* 43–50.

Rowell, J.A., Dawson, C.J. & Lyndon, H. (1990). Changing misconceptions: A challenge to science educators. *International Journal of Science Education, 12, 2,* 167–175.

Rubin, D. (2000). *Teaching elementary language arts: A balanced approach.* Boston: Allyn & Bacon.

Sabornie, E.J. & deBettencourt, L.U. (2004). *Teaching students with mild and high-incidence disabilities at the secondary level* (2nd edn). Upper Saddle River, NJ: Merrill-Prentice Hall.

Sadoski, M., Willson, V.L., Holocomb, A. & Boulware-Gooden, R. (2005). Verbal and nonverbal predictors of spelling performance. *Journal of Literacy Research, 36, 4,* 461–478.

San Mateo County Office of Education (2001). *Why do project-based learning?* Redwood City, CA: Project-Based Learning with Multimedia/San Mateo County Office of Education. Viewed 11 February 2006, (http://pblmm.k12.ca.us/PBLGuide/WhyPBL.html).

Scardamalia, M., Bereiter, C. & Steinbach, R. (1984). Teachability of the reflective process in written composition. *Cognitive Science, 8,* 173–190.

Schmedding, P. (2001). *About teaching science and technology (the 'thin edge' of science).* O'Connor, ACT: Parenting for a New Age. Viewed 11 February 2006, (http://www.geocities.com/horst1925/teachs.html).

Schonell, F.J. (1949). *Diagnosis of individual difficulties in arithmetic.* Edinburgh: Oliver & Boyd.

Schug, M.C., Tarver, S.G. & Western, R.D. (2001). Direct instruction and the teaching of early reading. *Wisconsin Policy Research Institute Report, 14, 2,* 1–31.

Schumm, J. & Vaughn, S. (1998). Issues related to instruction of students with learning disabilities. *Learning Disabilities Quarterly, 21,* 3–5.

SCoTENS (Standing Conference on Teacher Education North and South) (2005). *Teaching strategies to help with special educational needs: Teaching pupils with SLD (dyslexia).* Armagh, Northern Ireland: SCoTENS. Viewed 11 February, (http://www.socsci.ulst.ac.uk/education/scte/sen/teaching/dyslexiastrat.html).

Scott, B.J., Vitale, M.R. & Masten, W.G. (1998). Implementing instructional adaptations for students with disabilities in inclusive classrooms. *Remedial and Special Education, 19, 2,* 106–119.

Scruggs, T.E. & Mastropieri, M.A. (2003). Science and social studies. In H.L. Swanson, K. Harris & S. Graham (eds) *Handbook of learning disabilities* (pp.364–379). New York: Guilford Press.

Scruggs, T.E., Mastropieri, M.A., Bakken, J.P. & Brigham, F.J. (1993). Reading versus doing: The relative effectiveness of textbook-based and inquiry-oriented approaches in science education. *Journal of Special Education, 27*, 1–15.

Seguin, E. (1866). *Idiocy and its treatment by the physiological method.* New York: William Wood.

Serna, L. & Patton, J.R. (1997a). Effective teaching practices. In E.A. Polloway & J.R. Patton (eds) *Strategies for teaching learners with special needs* (6th edn, pp.135–173). Upper Saddle River, NJ: Merrill.

Serna, L. & Patton, J.R. (1997b). Mathematics. In E.A. Polloway & J.R. Patton (eds) *Strategies for teaching learners with special needs* (6th edn, pp.312–362). Upper Saddle River, NJ: Merrill-Prentice Hall.

SESD (Science Education for Students with Disabilities) (2006). New York: SESD. Viewed 3 February 2006, (http://www.sesd.info/index.htm).

Shenkman, H.L. (2002). Reading, learning, and thinking seminars: a template for faculty training. *Learning Abstracts, 5, 1*, 1. Viewed 11 February 2006, (http://www.league.org/publication/abstracts/learning/lelabs0201.html).

Shulman, L.S. (1987). Knowledge and teaching: Foundations of the new reform. *Harvard Educational Review, 57, 1*, 1–22.

Shupe, A.J. (2003). *Cooperative learning versus direct instruction: Which type of instruction produces greater understanding of fractions with fourth graders?* MA thesis. Morgantown, WV: West Virginia University.

Shuttleworth, V. (2000). *The special educational needs coordinator.* Harlow, UK: Pearson Educational.

Silbert, J., Carnine, D. & Stein, M. (1990). *Direct instruction mathematics* (2nd edn). Columbus, OH: Merrill.

Simon, C. (2001). Teacher assistant, or assistant teacher? *Special Children, 140*, 38–39.

Slavin, R.E. (1995). *Cooperative learning: theory, research, and practice* (2nd edn). Boston: Allyn & Bacon.

Slavin, R.E. (1999). Comprehensive approaches to cooperative learning. *Theory into Practice, 38, 2*, 74–79.

Smith, L. & Smith, D. (1997). Social studies topics, process and strategies. In J. Choate (ed.) *Successful inclusive teaching* (pp.336–366). Boston: Allyn & Bacon.

Solomon, G. (2003). Project-based learning: A primer. *TechLearning 23, 6*. Viewed 11 February 2006, (http://www.techlearning.com/db_area/archives/TL/2003/01/project.html).

Sotto, E. (1994). *When teaching becomes learning.* London: Cassell.

Special Connections (2005). *An introduction to classwide peer tutoring.* Lawrence, KS: Special Connections/University of Kansas. Viewed 1 February 2006, (http://www.specialconnections.ku.edu/cgi-bin/cgiwrap/specconn/main.php?cat=instruction&section=main&subsection=cwpt/main).

Staffordshire Learning Net (2005). *Special Needs Geography.* Stafford, UK: QLS/Staffordshire County Council. Viewed 3 February 2006, (http://www.sln.org.uk/geography/segsmain.htm).

Staley, D.J. (2005). From multimedia to multisensory education. *Threshold*, Fall issue, 28–31. Viewed 11 February 2006, (http://www.ciconline.com/AboutCIC/Publications/Archives/threshold_fall05.htm).

Stanford University (2001). Problem-based learning. *Speaking of Teaching, 11, 1*, 1–7. Viewed 11 February 2006, (http://ctl.stanford.edu/Newsletter/problem_based_learning.pdf).

Steel, B. & Hattersley, J. (2005). *GTIP Think Piece: Special educational needs.* Sheffield, UK: Geographical Association. Viewed 11 February 2006, (http://www.geography.org.uk/projects/gtip/thinkpieces/sen).

Stigler, J.W. & Hiebert, J. (1997). Understanding and improving classroom mathematics instruction. *Phi Delta Kappan, 79, 1,* 14–21.

Stigler, J.W. & Hiebert, J. (1999). *The teaching gap.* New York: Free Press.

Swanson, H.L. (2000a). What instruction works for students with learning disabilities? In R. Gersten, E. Schiller & S. Vaughn (eds) *Contemporary special education research* (pp.1–30). Mahwah, NJ: Erlbaum.

Swanson, H.L. (2000b). Searching for the best cognitive model for instructing students with learning disabilities: A component and composite analysis. *Educational and Child Psychology 17, 3,* 101–121.

Swanson, H.L. (2001). Searching for the best model for instructing students with learning disabilities. *Focus on Exceptional Children, 34, 2,* 1–15.

Swift, D. (ed.) (2005). *Meeting SEN in the curriculum: Geography.* London: Fulton.

Taylor, M. (2004). Geography teachers urged to inspire pupils: Inspectors say focus on facts alone fails to convey subject's relevance. *The Guardian,* 25 November 2004, 13.

Temple, C.M. (2001). Developmental dyscalculia. In S.J. Segalowitz & I. Rapin (eds) *Handbook of neuropsychology* (vol.7, pp.211–222). Amsterdam: Elsevier.

Tennant, G. (2001). The rhetoric and reality of learning support in the classroom: Towards a synthesis. *Support for Learning, 16, 4,* 184–188.

Thompson, H.J. (1962). *Colour factor mathematics.* Oxford: Heinemann.

Tobias, S. (1993). *Overcoming math anxiety.* New York: Norton.

Tomlinson, C.A. (2001). *How to differentiate instruction in mixed-ability classrooms* (2nd edn). Alexandria, VA: Association for Supervision and Curriculum Development.

Tompkins, G.E. (2006). *Literacy for the 21st century: A balanced approach* (4th edn). Upper Saddle River, NJ: Pearson-Merrill-Prentice Hall.

Topping, K. (1995). *Paired reading, spelling and writing.* London: Cassell.

Topping, K., Nixon, J., Sutherland, J. & Yarrow, F. (2000). Paired writing: A framework for effective collaboration. *Reading, 34, 2,* 79–88.

Tournarki, N. (2003). The differential effects of teaching addition through strategy instruction versus drill and practice to students with and without learning disabilities. *Journal of Learning Disabilities, 36, 5,* 449–458.

Tretiakov, A., Kinshuk & Tretiakov, T. (2003). Designing multimedia support for situated learning. In V. Devedzic, J.M. Spector, D. Sampson & Kinshuk (eds) *Proceedings of the 3rd IEEE International Conference on Advanced Learning Technologies* (pp.32–36). Los Alamitos, CA: IEEE Computer Society. Viewed 11 February 2006, (http://infosys.massey.ac.nz/~kinshuk/papers/icalt2003_alexei3.pdf).

TTA (Teacher Training Agency) (UK) (1998). *National standards for special educational needs coordinators.* London: TTA.

TTA (Teacher Training Agency) (UK) (1999). *National special educational needs specialists standards.* London, TTA.

TTA (Teacher Training Agency) (UK) (2000). *Using the national standards for special educational needs co-ordinators (SENCOs).* London: TTA.

Tunmer, W., Chapman, J. Ryan, H. & Prochnow, J. (1998). The importance of providing beginning readers with explicit training in phonological processing skills. *Australian Journal of Learning Disabilities 3, 2,* 4–14.

Tuovinen, J.E. & Sweller, J. (1999). A comparison of cognitive load associated with discovery learning and worked examples. *Journal of Educational Psychology, 91, 2,* 334–341.

Tweed, A. (2004). *Direct instruction: Is it the most effective science teaching strategy?* Arlington, VA: National Science Teachers Association. Viewed 11 February 2006, (http:// www.nsta.org/main/news/stories/education_story.php?news_story_ID= 50045).

Updike, M. & Freeze, R. (2002). Precision reading: Improving reading for students with learning disabilities. *International Journal of Disability, Community and Rehabilitation, 1, 1.* Viewed 11 February 2006, (http://www.ijdcr.ca/VOL01_01_CAN/articles/ updike.shtml).

US Department of Education, Office of Intergovernmental and Interagency Affairs (2004). *Helping Your Child Learn History* (2nd edn). Washington DC: US Department of Education. Viewed 3 February 2006, (www.ed.gov/parents/academic/help/history/ index.html).

Van Kraayenoord, C. (2002). Focus on literacy. In A. Ashman & J. Elkins (eds) *Educating children with diverse abilities* (pp.388–435). Sydney: Prentice Hall.

Van Kraayenoord, C. (2004). Teaching strategies for reading: How can we assist students with learning difficulties? In B.A. Knight & W. Scott (eds) *Learning difficulties: Multiple perspectives* (pp.67–84). Frenchs Forest: Pearson Education.

Van Kraayenoord, C. & Elkins, J. (1998). Learning difficulties in the regular classroom. In A. Ashman & J. Elkins (eds) *Educating children with special needs* (3rd edn, pp.131–176). New York: Prentice Hall.

Versey, J. (1993). *Differentiation: Managing differentiated learning and assessment in the National Curriculum (science).* Hatfield, UK: Association of Science Education.

Vincini, P. (2003). The nature of situated learning. *Innovations in Learning,* February issue. Somerville, MA: Tufts University. Viewed 1 February 2006, (http://at.tccs.tufts.edu/pdf/newsletter_feb_2003.pdf).

Von Heyking, A. (2004). Historical thinking in the elementary years: A review of current research. *Canadian Social Studies, 39,* 1. Viewed 11 February 2006, (http://www.quasar.ualberta.ca/css/Css_39_1/ARheyking_historical_thinking_ current_research.html).

Vygotsky, L.S. (1978). *Mind in society.* Cambridge, MA: Harvard University Press.

Wakefield, A.P. (1997). Supporting math thinking. *Phi Delta Kappan, 79, 3,* 233–236.

Wallace, T., Shin, J., Bartholomay, T. & Stahl, B. (2001). Knowledge and skills for teachers supervising the work of paraprofessionals. *Exceptional Children, 67, 4,* 520–533.

Waterfield, J. (2002). Dyslexia: Implications for learning, teaching and support. *Planet, Special Edition 3,* 22–23.

Watkins, C. & Slocum, T. (2004). The components of Direct Instruction. In N.E. Marchand-Martella, T.A. Slocum & R.C. Martella (eds) *Introduction to Direct Instruction* (pp.28–65). Boston: Allyn & Bacon.

Watson, B. & Kopnicek, R. (1990). Teaching for conceptual change: Confronting children's experience. *Phi Delta Kappan 7, 9,* 680–684.

Watson, J. & Boman, P. (2005). Mainstream students with learning difficulties: Failing and underachieving in the secondary school. *Australian Journal of Learning Disabilities, 10, 2,* 43–49.

Weiss, I.R., Pasley, J.D., Smith, P.S., Banilower, E.R. & Heck, D.J. (2003). *Looking inside the classroom: A study of K–12 mathematics and science education in the United States.* Chapel Hill, NC: Horizon Research Inc.

Weiten, W. (2001). *Psychology: Themes and variations* (5th edn). Belmont, CA: Wadsworth-Thomson.

West Virginia University (2005). *Disabilities, teaching strategies, and resources.* Morgantown, WV: Inclusion in Science Education for Students with Disabilities, West Virginia University. Viewed 11 February 2006, (http://www.as.wvu.edu/~scidis/sitemap. html).

Westwood, P. (2000). *Numeracy and learning difficulties.* Melbourne: Australian Council for Educational Research.

Westwood, P. (2001). *Reading and learning difficulties.* Melbourne: Australian Council for Educational Research.

Westwood, P. (2002). Are we making teaching too difficult? A critical look at differentiation in the classroom. *Hong Kong Special Education Forum 5, 1,* 13–29.

Westwood, P. (2003a). *Commonsense methods for children with special educational needs* (4th edn). London: Routledge-Falmer.

Westwood, P. (2003b). Drilling basic number facts: Should we or should we not? *Australian Journal of Learning Disabilities, 8, 4,* 12–18.

Westwood, P. (2004). *Learning and learning difficulties.* Melbourne: Australian Council for Educational Research.

Westwood, P. (2005). *Spelling: Approaches to teaching and assessment* (2nd edn). Melbourne: Australian Council for Educational Research.

Westwood, P. & Graham, L. (2000). Collaborative consultation as a component of support for students with special needs in inclusive settings: Perspectives from teachers in South Australia and New South Wales. *Special Education Perspectives, 9, 2,* 13–26.

White, S. (2005). Education that works in the Milwaukee public schools: Benefits from phonics and direct instruction. *Wisconsin Policy Research Institute Report, 18, 4,* 1–23.

Wilen, W., Ishler, M., Hutchinson, J. & Kindsvatter, R. (2000). *Dynamics of effective teaching* (4th edn). New York: Longman.

Willis, J. (1996). *A framework for task-based learning.* Harlow, UK: Longman.

Wong, B.Y.L. (2000). Writing strategies instruction for expository essays for adolescents with and without disabilities. *Topics in Language Disorders, 20,* 29–44.

Wong, B.Y.L., Butler, D.L., Ficzere, S. & Kuperis, S. (1996). Teaching low achievers and students with learning disabilities to plan, write and revise opinion essays. *Journal of Learning Disabilities, 29, 2,* 197–212.

Worthy, J., Broaddus, K. & Ivey, G. (2001). *Pathways to independence: Reading, writing and learning in Grades 3–8.* New York: Guilford Press.

Wragg, E.C. & Brown, G. (2001a). *Questioning in the primary school.* London: Routledge-Falmer.

Wragg, E.C. & Brown, G. (2001b). *Questioning in the secondary school.* London: Routledge-Falmer.

Yamanashi, J.E. (2005). A cooperative approach to assisting students at risk of educational failure. *Special Education Perspectives, 14, 2,* 62–76.

Yates, G.C.R. (1988). Classroom research into effective teaching. *Australian Journal of Remedial Education, 20, 1,* 4-9.

Yuen, M.T., Westwood, P.S. & Wong, G. (2005). Meeting the needs of students with specific learning difficulties in the mainstream education system: Data from primary school teachers in Hong Kong. *International Journal of Special Education, 20, 1,* 67–76.

Zhou, L. (2001). *Bridge the gap: Reflections on whole-class interactive teaching.* Exeter, UK: University of Exeter. Viewed 11 February 2006, (http://www.people.ex.ac.uk/PErnest/pome17/bridge.htm)

Zohar, A. & Aharon-Kravetsky, S. (2005). Exploring the effects of cognitive conflict and direct teaching for students of different academic levels. *Journal of Research in Science Teaching, 42, 7,* 855.

Index

strategy training: 49, 72, 90
  in mathematics **87, 90–92**
  in reading 72–73
  for writing 75, **76–77**, 78
structural apparatus 83
structured approach 3, 11, 15, 17, 18, **27, 30,** 70, 74, 85, **97**, 99, **102**, 109, 116, 120
student-centred methods **5, 11, 29–46**, 83, 116
students with disabilities 25, 61, 95, **101,** 110, 115, 119, 125, 128, 132, 133, 137
Studies of Society and the Environment 111
support: **127–138**
  in-class support 9, 57–58, 128, 129, **130–131**
  for learning 43, 58, 88, **119**, 125, **127–138**
  for teachers 127, 130, 136
support systems in schools 128, 129, 131, 135–136, 137
support teachers: duties **128–130**, 138
symbolic representation 103, **108**, 123
synthetic phonics **69–70**

T

task-approach strategies 5, **41–42**, 44
task-based language learning (TBLL) **39–41**
task-based learning (TBL) 39, 48–49, 96
teacher aides **128**
teacher assistance team (TAT) **135–136**, 138
teacher-directed methods 8, 9, **11–22**, 51, 60, 66, 72, 74, 76, 84, 92, 110
teachers:
  effective **5–6**
  expertise of **6–7**, 62
  personal qualities 6
  rapport with students 63
  subject knowledge of 4, **6, 13**, 88, **109, 117**
teaching:
  differentiating 3, **7–8**, 10, 106, **119**, 129
  effective instruction **5–6, 12–19, 86–87**
  poor quality **3–4**, 87–88, 107, **109, 117–118**
teaching adaptively **7–9**, 14, 95
teaching approaches:
  in environmental education 112, 114, 115, 124, 125
  in geography 114, 115–116, 118–119, 124, 125
  in history 122, **123, 124**
  in social studies 120–121
  as source of difficulty 1, **3–4, 14**, 15, **18,** 20, 21, **22, 26**, 27, **34, 36, 38**, 41, **42–43**, 44, **46**, 47, **59, 62**, 73, **88, 117**, 120

team teaching 119, 130
teamworking skills:
  of students 38
  of teachers 130, **135–137**
textbook approach to teaching 87, 98, 102, 106
textbooks 14, 25, 39, 56, 72, 96, 102, 105, **122**
thinking aloud: as a teaching strategy **42**, 44, **76**
thoughtfulness 112–113, 117, 118
time for learning 21
time management:
  by students 42, 109
  by teachers 3, 56, 118
transmissionist viewpoint 12
tutoring **57–60**, 90, 128

V

VAKT approach **26–27**, 70
visual aids 3, **13**, 88
visual discrimination 67
visual imagery 70, 77, 78
visual memory 2, 67
visual perception 2
vocabulary controlled books **66**
vocabulary knowledge: 14, 71, 104, 115, 121, 122
  influence on reading **71, 122**
volunteer helpers in school 128, **133–134**

W

wait time: when questioning **55**, 56, 104, 105, 121
whole-class teaching **8, 14–15**, 54, 58, 118
whole-language approach to reading 18, **66,** 74, 84
whole-math approach **84**
whole-word method:
  for reading **67**
  for spelling **77–78**
withdrawal model of support **129**, 130
word recognition **66**, 67, 69, 71, 105, 122
word study **71**
working memory 38, 71, 89, 90
worksheets 74, 76, 98, 105, 106, 107, 117, **119**
writing: 14, 16, 19, 65, 67, **73–77**, 116
  difficulties with **73–74**, 132
  teaching of 19, 41, 44, **74–75**

Z

zone of proximal development 43

NEW EDITION

Spelling

Approaches to Teaching and Assessment
Second Edition

Peter Westwood
ACER Press 1999, 2005

The new edition of **Spelling: Approaches to Teaching and Assessment** has been completely revised and expanded, with additional information on recent research-based teaching methods, many new references, online resources for teachers, and a new standardised spelling test for Australian schools.

New Edition Features

▮ Revised, updated and expanded, with new teaching methods and recent research findings

▮ New Form B for the **South Australian Spelling Test (SAST)**

▮ New normative data

▮ Updated resources section, including recent publications and online resources

Spelling provides an overview of some of the effective ways of helping students develop and improve their spelling skills. The emphasis throughout the text is on the importance of explicit teaching.

Many students need to be taught appropriate strategies for word study, and for editing and checking their own spelling. Research has supported the view that spelling skills can be improved by carefully structured intervention.

Because assessment is seen as an essential aspect of the effective teaching of spelling, **Spelling** also includes the **South Australian Spelling Test (SAST)** with new normative data, a glossary of 125 of the most commonly used words in children's writing, and diagnostic tests.

The formal and informal approaches to assessment described in the second half of this text will enable teachers to determine the instructional needs of individuals.

	0 86431 4124	$24.95
Peter Westwood set (5 titles)	A947BK	$140.00

Learning and Learning Difficulties

A Handbook for Teachers

Peter Westwood
ACER Press 2004

This outstanding teacher resource from bestselling author Peter Westwood explores a variety of learning processes, theories and concepts in order to help educators better understand and distinguish between the causes and outcomes of student learning problems.

Westwood aims to show that problems in learning are not all due to weaknesses within students or their lack of motivation. Two of the most powerful influences in the learning environment are the school curriculum and approaches to teaching. Westwood argues that many learning problems can be prevented or minimised by matching teaching methods and lesson content to a learner's current aptitude and prior experience.

	0 86431 7697	$34.95
Peter Westwood set (5 titles)	A947BK	$140.00

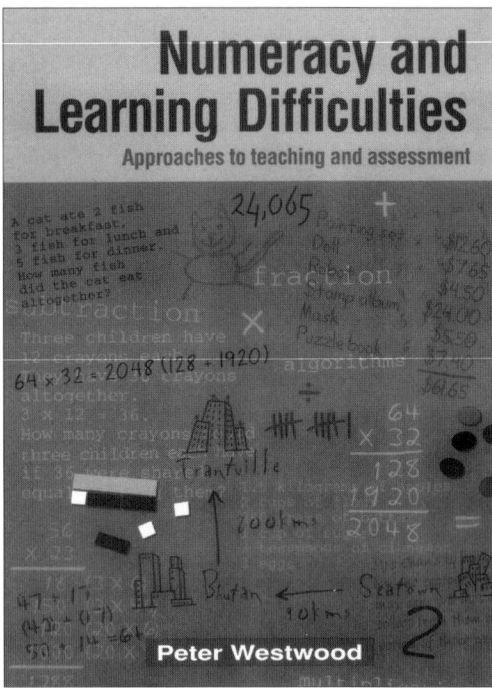

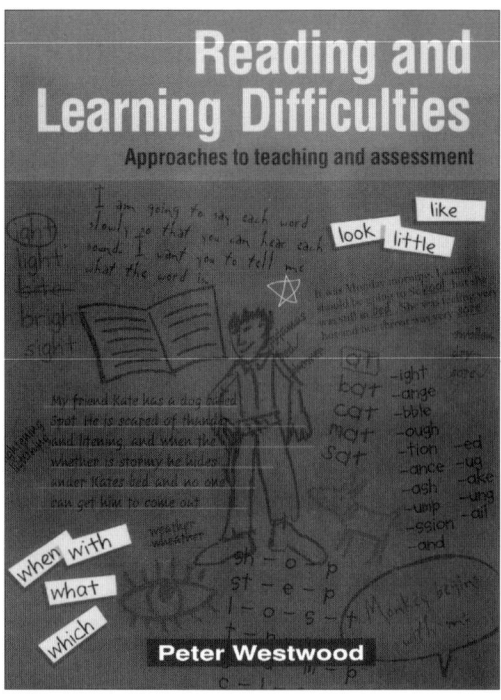

Numeracy and Learning Difficulties

Approaches to Teaching and Assessment

Peter Westwood
ACER Press 2000

This comprehensive guide examines the different ways students acquire mathematical skills, and helps teachers develop flexible teaching methods to suit these varied ways of learning.

Numeracy and Learning Difficulties also discusses common areas of learning difficulty in mathematics and 'why students fail'. It looks at ways teachers can determine gaps in students' knowledge, how to develop curricula to address these gaps, and problem-solving strategies and skills as a means of improving numerical literacy.

Drawing on research from the fields of developmental and cognitive psychology, Peter Westwood presents a case for high-quality 'first teaching' to prevent students failing in the initial acquisition of numeracy skills.

	A850BK	$29.95
Peter Westwood set (5 titles)	A947BK	$140.00

Reading and Learning Difficulties

Approaches to Teaching and Assessment

Peter Westwood
ACER Press 2000

This comprehensive guide to teaching reading more effectively presents a variety of research-supported approaches to teaching. These approaches have been designed to make learning to read easier and more successful for all children.

By examining the way readers process texts and identifying the knowledge and skills needed to become a proficient reader, author Peter Westwood explains why learning problems can sometimes occur and what can be done to prevent or overcome these difficulties.

	A945BK	$29.95
Peter Westwood set (5 titles)	A947BK	$140.00

Melbourne

19 Prospect Hill Road
CAMBERWELL
Victoria 3124

Tel: (03) 9277 5555
Fax: (03) 9277 5500
<www.acer.edu.au>

Sydney

1/140 Bourke Road
ALEXANDRIA
New South Wales 2015

Tel: (02) 8338 6800
Fax: (02) 9693 5844
<www.acer.edu.au>

Brisbane

9 Hill House
541 Boundary Street
SPRING HILL
Queensland 4000

Tel: (07) 3831 2769
Fax: (07) 3831 9900
<www.acer.edu.au>

ACER Press

347 Camberwell Road
CAMBERWELL
Victoria 3124

Tel: 1800 338 402 (Toll free)
(03) 9835 7447
Fax: (03) 9835 7499
Email: sales@acer.edu.au
<www.acerpress.com.au>

For information about products and services please contact **ACER Press**.

ACER